D0436182

THE NETWORK

THE NETWORK

The Battle for the Airwaves and the Birth of the Communications Age

SCOTT WOOLLEY

An Imprint of HarperCollins*Publishers*

 For information address HarperCollins Publishers, 195 Broadway, New York, NY 10007.

HarperCollins books may be purchased for educational, business, or sales promotional use. For information please e-mail the Special Markets Department at SPsales@harpercollins.com.

FIRST EDITION

Designed by Suet Yee Chong

Library of Congress Cataloging-in-Publication Data has been applied for.

ISBN 978-0-06-224275-4

16 17 18 19 20 OV/RRD 10 9 8 7 6 5 4 3 2 1

AUTHOR'S NOTE

The story that follows is true.

The characters' dialogue represents their exact words as captured by reliable contemporaneous recordings, including audiotapes, courtroom transcripts, and police wiretaps.

CONTENTS

ACT III

INFINITE

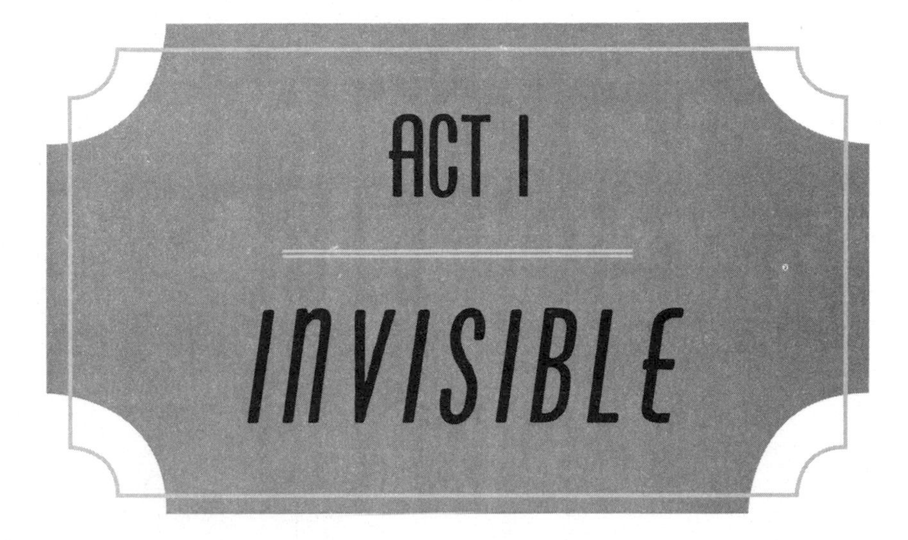

ACT I

INVISIBLE

CHAPTER 1

ARMSTRONG

New York City—1954

EDWIN ARMSTRONG HAD HIS THIRTEENTH-FLOOR LUXURY apartment all to himself for the night, just as he had planned. His wife, Marion, was staying at her sister's in Connecticut. He had surprised their live-in cook and two maids with an extra night off work. The world-famous inventor, Ivy League professor, and multimillionare placed his handwritten suicide note on the bedroom dresser and walked to the window. He slid it open, feeling a shock of freezing winter air wash over him.

The Armstrongs' sprawling seventeen-room apartment was the sort of place most New Yorkers dreamed of living. Known as River House for its location on the western bank of the East River, the twenty-six-story art deco tower had been designed in the Roaring Twenties to cater to the top rung of Manhattan's social ladder. Its other fifty-five apartments housed neighbors with last names like Roosevelt, Hearst, and Rockefeller. The Armstrong apartment, set at the midpoint of the building's A line, offered one of River House's signature views: a panoramic sweep from Brooklyn to Queens, across Roosevelt Island, and up the Manhattan skyline running north along the water's edge. In past summers, the river had served as convenient parking spot for his neighbors' yachts. Tonight, a fierce wind blew upriver, speckling the dark water with whitecaps.

To his fellow New Yorkers, the day just ended had been an ordinary Sunday—but to Major Armstrong it marked an important anniversary. Forty years ago, he had witnessed the full power of his first invention, a discovery that kicked off his career as the most prolific inventor since Thomas Edison.

On that long-ago January 31st, in 1914, a twenty-three-year-old Edwin Armstrong had embarked on an unusual adventure. Back then, the ability to send information through the air was little more than a parlor trick. Consumer radios had not yet been built, or even imagined. The only thing the airwaves were good for at the beginning of the twentieth century was zipping Morse code dots and dashes a few dozen miles—useful for communicating with ships at sea but little else. Armstrong believed he had invented a device that could change that, grabbing enervated airwaves drifting in from clear across the Atlantic and reinfusing them with enough power to make them detectable, a feat that wireless experts considered impossible.

That's when David Sarnoff first entered the story. An acquaintance of Armstrong's, Sarnoff shared the inventor's youth and optimism, and was one of the few people to believe Armstrong's idea might actually be possible. Since Sarnoff worked for a company that owned the world's largest antenna, he offered to connect Armstrong's device to the antenna and put his claims to the test. The results shocked even the two young optimists as they suddenly began to pick up wireless signals being sent from the far side of the planet.

From that day on, Armstrong and Sarnoff's fascination with the untapped power of the invisible waves would bond them and set their lives on parallel paths. Armstrong's imagination, talent, and luck continued to astound. Following the creation of his tool to make weak airwaves stronger, he built a device that made it easy for a radio transmitter to summon the invisible waves. He followed that with a breakthrough that allowed engineers to change an airwave's length, multiplying their information-moving power once again. Invisible waves controlled by Armstrong's discoveries would deliver the sound of Duke Ellington's jazz to radio listeners, the sight of Jackie Gleason's antics to TV viewers, and the location of incoming Nazi bombers to the Allied Air Command.

Like Thomas Edison, who kept working into his eighties, Major Armstrong had maintained his passion and creativity as he had aged. Twenty years after their trip to test out his first invention, Armstrong invited Sarnoff to his lab at Columbia University to show him yet another invention, FM radio.

EDWIN ARMSTRONG'S BIGGEST MISTAKE, as he now saw it, had been his naive assumption that the David Sarnoff who showed up for the FM radio demonstration, would be the same giddy engineer who had accompanied him twenty years earlier. He knew, of course, that his friend had used their shared faith in the airwaves' expanding power to build a wildly successful career of his own, ascending to run the Radio Corporation of America, the world's largest manufacturer of AM radios and other electronics, as well as the National Broadcasting Company, RCA's most famous subsidiary and the world's largest AM radio network. What Armstrong hadn't expected was for David Sarnoff to betray their friendship and lead a secret cartel dedicated to crippling both FM radio and the infant television industry.

When Armstrong first turned his efforts to convincing the world of those explosive allegations, following the end of World War II, he knew it would be a challenge. By then, David Sarnoff was, in the American public's eyes, famous, accomplished, and respected. Even so, Armstrong's task of exposing the corruption and Sarnoff's role in it hadn't seemed so daunting at first. After all, the inventor had spent his life revealing invisible powers to an unbelieving public.

And yet, for reasons he still did not understand, nothing worked. Not the seventeen lawyers he hired to sue Sarnoff and RCA. Not the congressional investigations he instigated. Not the accusations he leveled in the popular and the scientific press. None of it ever seemed to matter.

Why had all of his efforts to expose the corruption of the American airwaves failed?

Who was the real David Sarnoff?

Those questions had obsessed the inventor for nearly a decade, his failure to answer them slowly poisoning his marriage, his career, his life. Now, as the biting wind blowing up the

East River washed over him and flooded into the apartment, he considered his options. The blueprints for River House put the distance to the rooftop below him at 139 feet, 7 inches. More than enough for the job, the engineering professor could easily calculate, just by eyeballing the drop.

CHAPTER 2

SARNOFF

New York City—1906

A HEAD OF CURLY BROWN HAIR BOBBED ALONG IN THE RIVER OF older, taller men rushing to work in lower Manhattan. At Wall Street the main flow of commuters forked off in the direction of the New York Stock Exchange while the fifteen-year-old continued south, down a side street.

David Sarnoff made for an unusual sight, and he knew it. Fifteen was young to be starting a full-time job on Wall Street, and he looked closer to twelve, thanks to his stubby stature and the baby fat filling out his round face. Sarnoff hoped the formal business suit he wore was doing its job of disguising his youth, though from the glances of his fellow commuters it seemed possible that it was having the opposite effect, drawing attention to the comic contrast between his age and his attire.

Yet the teenager gave off an air of nonchalance. It was one of the immigrant's peculiar talents: a knack for fitting in at whatever new world he found himself. Part of that gift was innate, a self-assurance rarely found in adults and invariably absent in undersized teenage boys. The other part of David Sarnoff's easy adaptability could be traced to hard-won experience. Born in 1891, Sarnoff spent the first nine years of his life in an isolated Jewish village, speaking Russian and Yiddish. Then, one day, his family shipped out for Manhattan, where he had spent the last six years learning English and everything else he could about his new home.

Even as a nine-year-old unable to comprehend the language, Sarnoff was immediately entranced by turn-of-the-century New York. Its four million people. Its electric streetlights. Even New York's clouds were exciting, thanks to local newspapers' habit of using them as floating billboards upon which giant spotlights blinked out election results, sports scores, and other big news in Morse code. But to David Sarnoff the real difference between his old village and his new city was more fundamental. In the shtetl where he spent the last decade of the nineteenth century, much of it studying the Torah, his daily life was almost indistinguishable from that of his seventeenth-century ancestors. In twentieth-century New York, stumbling across the unimaginable became a regular and delightful part of David Sarnoff's life.

His fellow commuters offered one example. When Sarnoff and his family arrived in 1900, he had been floored by the city's extensive network of elevated steam trains and horse-drawn trolleys. Today, just six years later, he could see his fellow commuters materializing on Wall Street, delivered to the financial district by the new subway system that used electric power to whisk them underground and eliminate the soot and manure of aboveground transportation. Commuters could now travel from Grand Central Station in midtown Manhattan to Wall Street in a mere twelve minutes.

To David Sarnoff, the subway's nickel fare remained a luxury, especially on a sunny September morning like this one. From the Lower East Side tenement he shared with the rest of his immigrant family, it took about fifteen minutes to walk to Wall Street. Besides, what really fascinated the fifteen-year-old immigrant was not mankind's newfound ability to move people beneath cities at the speed of a steam train, but its power to move information around the globe at the speed of light.

THE TEENAGER HAD HAPPENED UPON his new obsession at the end of another difficult path. Ever since arriving in New York, Sarnoff's father, Abraham, had been incapacitated by tuberculosis. David helped support his family by rising before dawn to manage his own newsstand before heading off to school. His father's death earlier this year ended his schooling, and made him the primary breadwinner for his mother, Leah, and two younger brothers. Having just completed eighth grade, Sarnoff declared his years of part-time work and full-time schooling over and began looking for a job.

After scouting around for a job as a newspaper copyboy, Sarnoff happened to stop by the headquarters of the Commercial Cable Company, which was hiring messenger boys. Suddenly and unexpectedly, the fifteen-year-old found himself in a dream job.

"The Commercial," as the company was known, was arguably 1906 New York's hottest high-technology company, if not the world's. Founded by mining baron John Mackay, the company owned five of the sixteen undersea cables that sent telegrams beneath the Atlantic Ocean. The Commercial also ran one of the two new cables that crossed the Pacific, a ten-thousand-mile-long wire that could zip text messages from California to Japan, China, and the Philippines.

Each cable was an expensive feat of engineering, paid for with silver dug from Mackay's mines in California and Nevada. The newest of the Commercial's five Atlantic cables, built for a record $3.5 million, contained 1.4 million pounds of copper insulated with 800,000 pounds of sap from Malaysian perch trees. Another 17 million pounds of armoring made from a blend of brass, iron, and jute protected the very expensive wire from the dangers of the deep: sharp rocks on the ocean floor, ships' anchors, and teeth of curious sharks attracted by the cable's electrical current.

For the new messenger boy, the connections made possible by this ever-expanding worldwide web of cables held a special fascination. At the beginning of the last century, the entire planet had been as isolated as his old Russian village. If two people wanted to exchange information instantly, they needed to move close enough to see or hear each other. Moving information across or between continents meant moving physical objects, usually people or paper, at low speed and high cost. The resulting delays meant that distant continents frequently experienced major events in isolation, as natural disasters, wars, and other crises often began and ended before the rest of the world heard the news. Back-and-forth conversations between distant lands could take months, if not years.

The first hint that a faster form of communication might be possible came in 1822, when Danish physicist Christian Órsted hooked a copper wire to a battery and noticed something odd about a magnetized compass needle that sat nearby. French scientist Louis Pasteur summarized the importance of what happened next: "He suddenly saw (by chance you will say, but chance only favours the mind which is prepared) the needle move and take up a position quite different from the one assigned to it by terrestrial magnetism. A wire carrying an electric current deviates a magnetized needle from its position. That, gentlemen, was the birth of the modern telegraph."

By the 1830s, thanks to the work of inventors led by William Cooke, Charles Wheatstone, and Samuel Morse, neighboring cities began using Órsted's discovery to trade instant messages. A sender in one city would send an electric current down a telegraph wire, making a magnet move at the far end. By using a "telegraph key" to interrupt the current, the tempo of dots and dashes tapped out on the key would be mirrored by the movements of the distant magnet. It took another three decades of experimenting—and the genius of the great Victorian scientist Lord Kelvin—before the first transatlantic cable made it possible to jiggle a magnet on the other side of an ocean in 1858. The job of crossing the much wider Pacific Ocean proved far more challenging. David Sarnoff had been in New York for three years when the newspapers he sold announced that on July 4, 1903, President Theodore Roosevelt had made history by sending himself a telegram that traveled clear around the earth.

In turn-of-the-century New York, David Sarnoff's fascination with the growing global network was unusual only in its degree. To most ordinary people, "the electrical transmission of intelligence" was as close as modern science had come to discovering magic. Former secretary of state Edward Everett captured the general sense of awe after the first Atlantic cable carried an inaugural text message from Queen Victoria to President Buchanan: "Does it not seem all but a miracle of art that the thoughts of living men—the thoughts that we think up here on the earth's surface, in the cheerful light of day—about the markets and the exchanges, and the seasons, and the elections, and the treaties and the wars, and all the fond nothings of daily life, should clothe themselves with elemental sparks and shoot with fiery speed, in a moment, in the twinkling of an eye, from hemisphere to hemisphere?"

With every milestone, newspapers competed to express the general sense of wonder at the ever-expanding, worldwide web of cables. "The magnitude of the subject transcends the power of language," the *New York Times* swooned after the first Atlantic cable went live. "The Atlantic has dried up and we become in reality, as well as wish, one country," declared *The Times* (London). The public was equally euphoric. One celebration in Manhattan following the first telegram from Europe got so out of hand that much of city hall ended up in ashes. After President Roosevelt's 1903 telegram made it around the earth, many commentators took a special glee in noting that twentieth-century science had outdone even William Shakespeare's Puck, the magic fairy who brags of being able to circle the earth in forty minutes. Teddy's telegram raced around the world three times that fast.

For David Sarnoff, an extrovert who grew up cut off from

the world, working in the Commercial Cable Company's office, connected to the far points of the globe, had felt like being admitted to the most magical building in the world's most magical city. It also had magnified the pain and humiliation he felt three weeks before, when his boss fired him for taking a day off work on Rosh Hashanah.

Today marked the first day of his second chance to make it in the global communications business. Unlike the respected, profitable, and well-established Commercial Cable Company, his new employer had never posted a profit and many people on Wall Street suspected the company of being a front for an elaborate stock fraud. The fifteen-year-old Sarnoff had no way of independently evaluating that gossip. Besides, thrown out of the cable industry, for the moment he had no choice but to cast his lot with the rebels. Leaving the cool fall air behind, David Sarnoff walked inside and began introducing himself around the Marconi Wireless Telegraph Company of America.

THE MARCONI COMPANY did have its believers. Optimists viewed the four-year-old company as a high-tech innovator, an upstart that had the potential to replace expensive wires under the sea with cheap waves in the air.

Though the company's new office boy possessed no real scientific training, the gear he saw being used to send wireless messages remained rudimentary enough for him to quickly pick up a basic understanding of how it worked. The first scientist to shoot dots and dashes through the empty air (instead of a copper wire) had accomplished the feat in 1887, four years before David Sarnoff was born. At the time, Professor Heinrich Hertz, a German academic, had hoped to impress his fellow physicists by confirming a theory dreamed up by Scottish physicist James Clerk Maxwell in the 1860s. Using fancy mathematics, Maxwell had predicted the existence of invisible and intangible waves made from equal parts electricity and magnetism, but until Hertz came along, no one had been able to prove that Maxwell's waves existed in the real world.

Hertz dreamed up a novel way to test the theory by mimicking a lightning bolt. While religion often attributed lightning to the divine ("God's burning finger," Herman Melville called it), Benjamin Franklin and other scientists studied it for clues to the true nature of electricity. Before Hertz's experiments, one of lightning's most intriguing and baffling powers was the way a bolt from the sky announced itself with a visible flash of lightning, an audible clap of thunder—and also in a third voice that almost always went unnoticed. Occasionally, lightning strikes caused metal objects to quiver or throw off tiny sparks at the exact moment lightning flashed in the distance. Professor Hertz saw this as a sign that the bolts were emitting the invisible electromagnetic waves Maxwell's equations predicted.

To re-create the invisible, intangible waves in his lab, Hertz hooked up a simple device that fed electricity into a piece of metal until the charge leaped through the air to a nearby piece of metal in the form of a visible, airborne spark. These miniature lightning bolts did indeed release a burst of electromagnetic waves, which Hertz detected on the far side of his lab using a simple wire loop. (Hertz found that a circle of wire, with a small gap on one side, would emit its own much smaller spark every time the spark appeared on the far side of his lab.) While Hertz's academic colleagues hailed his achievement, the rest of the world paid it no notice. As even Professor Hertz acknowledged, waves that no one could see or touch were obviously "of no use whatsoever."

Fifteen years later, the wireless telegraph transmitters that David Sarnoff saw around the Marconi Company were more powerful but otherwise identical versions of Hertz's original "spark gap" wave generator. The messenger boy watched as Marconi operators tapped out a message using a standard telegraph key, each touch summoning a spark in the air. To make dots and dashes the wireless operator would press the key in a staccato pattern, summoning a spark that disappeared almost instantly (a dot) or one that hung in the air a split second longer (a dash).

The equipment Sarnoff found the operators using to catch the waves represented a vast improvement over the wire loop Professor Hertz used as his original receiver. New magnetic detectors nicknamed "Maggies" took advantage of the basic relationship between electricity and magnets Hans Órsted discovered nearly a century earlier. An incoming electromagnetic wave would be caught on an antenna, producing a faint electrical current that moved a magnet. The Maggies did an excellent job of taking those tiny movements and using them to switch an audible signal off and on. A Marconi operator could sit with a pair of headphones plugged into the Maggie and hear the Morse code tapped out by a distant operator.

Despite the steady advance in wireless technology, when

skeptics looked at the Marconi Wireless Telegraph Company of America they saw a company designed not to connect continents with each other but to separate investors from their money. While no one questioned the ability of spark gaps and Maggies to send messages a few hundred miles, the company's promises to offer a commercial service sending telegrams all the way to Europe had yet to materialize. The skeptics' case had been strengthened by the recent exposure of several of Marconi's smaller competitors as precisely that, fraudulent companies that sold stock based on wireless technology that turned out not to exist. Stock in Sarnoff's employer, which had been sold to the public in 1902 for sixty-seven dollars a share, now traded in the midthirties.

In the first few weeks of his new job, David Sarnoff turned up a few clues that only deepened the mysteries of the Marconi Company. One of the first things to jump out at the new office boy was the odd quiet that pervaded the company offices every morning. At the Commercial Cable Company, the workday started with a bang. At exactly 10 A.M., the New York Stock Exchange opened and a flurry of buy and sell orders began zipping beneath the ocean, a frenzy that lasted for exactly one hour—even though the New York Stock Exchange remained open for another five.

This part of the business was no puzzle to Sarnoff. While ordinary people imagined the world's undersea telegraph cables ferrying news reports, diplomatic proposals, and military orders, his days as a messenger boy had acquainted him with the cables' main customers and their peculiar business. Since the first transatlantic telegraph began offering commercial service in 1865, the undersea cables had created an industry built on the newfound ability to spot identical goods selling for different prices on different continents. "International arbitrageurs" would buy shares of stock in London and then, a few minutes or a few seconds later, sell the same shares for a bit more in New York. When the price gap swung in the other direction, so did their trades—but either way a profit was guaranteed as long as both trades could be quickly executed before the price difference disappeared. In London, the hectic hour of stock arbitrage began at 3 P.M. (when the New York Stock Exchange opened) and ended an hour later, when the London exchange closed for the day. After that, cable traffic dropped off significantly, though messages continued to ping between arbitrage firms buying and selling bars of gold, bales of cotton, bonds issued by railroads, and any other type of easily traded asset they could find selling for different prices on opposite sides of the Atlantic.

Although Sarnoff's bosses often spoke of the Marconi Company as a cable rival, the new office boy saw no Marconi telegraph operators sending buy and sell orders across the ocean—nor any other type of transatlantic messages, for that matter. A trip to the Marconi Company's newest wireless station, on Coney Island, was only slightly more reassuring. The station had wireless operators, but from what Sarnoff could see, their days were painfully slow. Every hour or so they would put on their headsets and begin tapping out brief text messages to a ship entering or leaving New York's harbor. As far as he could tell, the only way they could send a message all the way to Europe was by hopscotching it across a series of ships—and that cumbersome process would only work when a group of Marconi-equipped ships happened to line up across the Atlantic.

WORK COULD BE UNPLEASANT for Sarnoff. Other Marconi employees had taken to calling the messenger boy "Davey" and, on occasion, "Jew boy." Sarnoff bore the insults silently, and resolved to work his way up to a position where he would not be forced to tolerate such derision.

Like many tech-minded teenagers, David Sarnoff dreamed of landing a job as a telegraph operator. Operators with a good "fist" could move forty words per minute, while a "copperplate fist" could transmit close to fifty. Sarnoff immediately devoted himself to improving his own sending speed, buying a used two-dollar telegraph key and practicing for hours a night after work. Experienced fists had an endless list of insulting nicknames for poky amateurs—"Rattle brain," "Jay," "Swell head," "Crank"—which Sarnoff was determined to never have stuck on him.

Around the main Manhattan office, Sarnoff found other company records that offered new clues about his controversial employer. John Bottomley, the chief executive of American Marconi, had taken to bragging that thanks to the new Seagate station in Coney Island, his company now offered an "unbroken string of stations" from New York Harbor to Newfoundland. Still, the ship-to-shore business those six stations did was disturbingly infrequent. According to the last numbers the company made public, the Marconi Company was sending an average of forty-one wireless telegrams per day, a total of just 557 billable words. Even that exaggerated the true volume of business, since American Marconi split the revenue from each message with the separate company that owned the wireless gear on ships, and often had to cut in the land-based telegraph companies whose wires carried the messages from the shore stations back to New York.

The company's latest financial statement avoided any mention of profit or loss, but it didn't take formal finance training for David Sarnoff to work out that the company had been burning cash since its founding. The company's primary source of income was selling its own stock, including $117,000 in 1903 and another $293,000 in 1904. The following year it hadn't sold any stock, allowing its operating losses to rapidly erode the company's remaining cash. The one bright spot on the Marconi Company balance sheet was its patent portfolio, which had been built up through the acquisition of patents from famous scientists with names like Edison, Lodge, and Fleming. If the company's accountants were to be believed, these patents were worth a rich $5.5 million.

The Marconi Company's cash shortage was an open secret. On several paydays his boss sent David racing across the city to pick up an emergency loan to keep the company's paychecks from bouncing.

DAVID SARNOFF'S GROWING curiosity about his new employer began to focus, naturally enough, on the mysterious character whose name was on the door but whom Sarnoff had never seen around the office.

Guglielmo Marconi—or at least the Guglielmo Marconi of David Sarnoff's imagination—was equal parts inventor, businessman, and showman. After a few months on the job, David Sarnoff caught wind of the Italian inventor's upcoming visit to New York. For the rest of his life, David Sarnoff would remember hiding out in the Marconi Company office, hoping for a moment alone with his new idol. The inventor seemed charmed, or at least bemused, at the sight of the small, bold messenger boy popping out from behind a row of equipment and introducing himself with all the seriousness he could muster.

CHAPTER 3

MARCONI

New York City—1912

FAINT WISPS OF SMOKE ROSE INTO THE DENSE MORNING FOG above Guglielmo Marconi's head, adding to his annoyance. The inventor needed to get to New York as soon as possible. He had switched his ticket to the RMS *Lusitania* from the RMS *Titanic* to catch the older ship's earlier departure. The *Lusitania,* scheduled to arrive in New York the previous day, already had been delayed by a fierce storm in the mid-Atlantic. Now the famous ocean liner floated idly at the mouth of New York Harbor, a wall of morning fog blocking its entrance. While the midmorning sun had begun to burn off the fog, it apparently was not enough for the captain, based on the lack of smoke coming out of the ocean liner's four giant funnels.

In between the *Lusitania*'s middle two smokestacks, which poked through the ship's sky deck, Marconi could see an unusual hut. There were countless ways for the ship's first-class passengers to kill time on board, including a library, smoking room, and a veranda café that in sunnier weather offered a chance to eat outdoors. An accomplished piano player, Marconi often entertained his fellow passengers with informal concerts. He also enjoyed coming here, to the hut from which the familiar clatter of a spark-gap transmitter emanated.

Like all experienced telegraph operators, Marconi was fluent enough in Morse code to translate the outgoing dots and dashes to words in his head. Usually, he paid little attention to the chatter. On the deck below, bored passengers often wrote what amounted to electronic postcards, making a typical day's traffic a never-ending stream of banal personal greetings. (A typical 1912 Marconigram: "Jane is fine and sends you a kiss. Everyone well, Grannie.")

Overhearing such greetings left the Italian inventor with mixed feelings. He owed his company's survival to the maritime market, which continued to be its only source of profit. Modern steamships could cross the Atlantic in less than a week—the *Lusitania*'s record time of four days, sixteen hours, and forty minutes had been beaten only by her sister ship, the *Mauretania*—but that was still longer than many passengers wanted to be out of touch. With the growing fleet of giant passenger liners crisscrossing the world's seas, Marconi's ship-to-shore business continued to expand steadily. Thanks to the maritime messages, the Marconi Wireless Telegraph Company of America had managed to report a paper-thin profit last year. Still, the Italian inventor viewed the business as he always had, as a placeholder—a niche business that would pay the bills until the transoceanic business took off and the real riches arrived.

Marconi enjoyed visiting the *Lusitiania*'s wireless shack even though the hut and its gear were embarrassingly rudimentary. The ship was only five years old, but back then shipbuilders didn't think to build a cabin for a wireless telegraph operator. If he hadn't been in such a hurry, Marconi would have waited to travel to New York aboard the *Titanic*. Newer ocean liners, including the about-to-launch *Titanic* and *Olympic,* came complete with all the latest technology. As a result, those newer ships could more than double the 215-mile range of the *Lusitania*'s wireless transmissions, a major leap forward that Marconi would have preferred to witness firsthand, had his trip to New York not been so urgent.

Finally, smoke began to pour out of the funnels and the ship began to move slowly into the harbor, the rat-a-tat of the telegraph sending word to the Marconi station at Seagate as well as the newer station in downtown Manhattan. Once inside the harbor the ship headed for Staten Island, where quarantine inspectors and an enterprising newspaper reporter awaited her arrival.

Ten years after Guglielmo Marconi first claimed to have sent a telegram across the Atlantic, he was once again returning to New York to ask American investors for money to build a new generation of wireless stations. Those stations, he believed, would allow his company to compete with the hyperprofitable transatlantic cables. It promised to be a tricky sales pitch, Marconi realized. His task had been made much harder by enterprising con artists who had seized on the public fascination with the technology to sell stock in phony wireless companies with names like the Pacific Wireless Telegraph Company and the Massie Wireless Telegraph Company. One financial reporter compiling a list of wireless frauds for a magazine counted "nearly a score, none of which, either singly or in combination, has developed a commercial success." The resulting article, run under the headline "WIRELESS & WORTHLESS" got right to the point: "More men are now in prison or under indictment for selling stock in wireless tele-phone and telegraph companies than is the case with any other line of industrial promotions."

"The very word 'wireless' brings a smile to the lips of the Wall Street man," warned another magazine exposé. "Thousands of men and women in this country have already learned to curse the day Marconi made his first experiment."

GUGLIELMO MARCONI FIRST STUMBLED into the wireless business a few years after Professor Hertz's famous 1887 experiment, which the young Italian first copied and then set about trying to improve. After a few years spent experimenting in his parents' attic, the then-unknown twenty-two-year-old showed up in London in 1896, carrying two mysterious boxes. Marconi put on a series of public demonstrations, amazing British audiences with a showier re-creation of Hertz's eight-year-old feat. First he would ask a volunteer to carry one of his magic boxes around the lecture hall. Onstage, Marconi would build tension. Then, with a flourish, he would push a lever protruding from the box he kept next to him, an action that immediately elicited a buzz from the second box and gasps from the audience. Astonished onlookers leapt to their feet in amazement. As a reporter marveled after one such performance: "A signal was being sent around the lecture hall that was invisible, but as tangible in its effects as any telegraph impulse sent along a wire."

On the back of such publicity, Marconi raised enough

money from British investors to hire the country's top patent attorney, file for a patent on his "system for wireless telegraphy," and found the world's first company devoted to wireless communication. While Marconi possessed indisputable talent as a businessman and promoter, future generations of scientists would give him credit for personally designing one relatively minor component of his patented system, a better way to connect the antenna that extended the distance the waves could travel and still be detected.

Using his investors' cash, Marconi staged a series of public demonstrations as he hunted for a wireless application that customers might be willing to pay to use. In 1898, he set up a telegraph on the prince of Wales's yacht so Queen Victoria could send him messages. ("Can you come to tea?" the queen texted. "Very sorry, cannot come to tea," replied the prince.) At the Kingstown Regatta, a popular race off the coast of Ireland, Marconi sent the winning boat's name to the *Dublin Daily Express,* letting the paper scoop its rivals. Eventually, however, Marconi's charm and knack for winning free publicity wore thin, as his fed-up investors demanded that he find a profitable application for wireless communications or close up shop and return what little remained of their money.

At the time, Marconi's odds of winning the race to send the first message over an ocean appeared slim. Several other rival inventors were racing to beat him. Nikola Tesla, the brilliant Serbian inventor, was nearing completion of his own massive transmitting tower on New York's Long Island. Bankrolled by James Pierpont Morgan, Tesla's tower raised a fifty-five-ton steel structure 187 feet high in the air, a sort of super-antenna of the type that the cash-strapped Marconi Company could never afford. Tesla, himself a showman who often amazed live audiences by shooting white bolts of lightning out of his fingertips, had boasted publicly of his plans to beat Marconi, vowing that his soon-to-be-completed tower would possess "a grip on the earth so the whole of this globe can quiver."

Marconi used the threat of Tesla to persuade his board of directors to invest the company's remaining cash on an all-out effort to win the race across the Atlantic. Many of the directors considered the effort a desperate gamble. After five years, British Marconi had plenty of experience in extending the range that its Morse code messages could travel. Thanks to Marconi's talent for integrating other inventors' discoveries into his company's wireless system, the company had managed to regularly extend the range of its messages. Barely able to make it a single mile in 1896, they managed an 8-mile hop in 1897, leaped across the English Channel in 1899, and in early 1901 set another new record of 225 miles. If Marconi managed to leap the Atlantic, it would be by far the biggest improvement yet, a distance of nearly 3,000 miles.

To make the jump, Marconi told his board in 1901, he just needed more powerful transmitting stations and larger receiving antennas, which they approved and he rushed into construction. The hastily constructed stations looked impressive: on each side of the Atlantic, a circle of twenty wooden masts rose two hundred feet into the air, suspending four hundred wires in the shape of a giant metal ice-scream cone. The tight budget and hurried schedule forced Marconi to skimp on structural engineering however, and within months, a storm knocked over the Marconi station in Britain. Two months later, a storm in Cape Cod left that antenna in splinters too.

At that point, even his staunchest backers recognized Marconi's dream was dead, at least for the foreseeable future. The only person not willing to quit was Guglielmo Marconi himself: he booked passage to Newfoundland before his board of directors could order him back to London.

The rocky hill on the Canadian coast where Marconi set about trying his final, last-ditch effort was a particularly miserable place in mid-December, wedged between the Atlantic Ocean and St. John's, Newfoundland, the windiest, most rain-soaked city in Canada. Coming to Newfoundland shortened the distance his signal would have to travel, but even the narrow neck of the North Atlantic was ten times as wide as the distance traveled by his last record-setting transmission. (A rough rule of thumb known in the telegraph industry as the "law of squares" magnified that challenge: a ten-fold increase in the distance traveled meant the signal would arrive not with one-tenth of the power of the previous record, but just one one-hundredth.)

And then there was the kite. Unable to afford a proper antenna tower, Marconi planned to use an oversized kite made of bamboo and linen to raise a wire into the air. The best antennas, he knew, are set at a fixed length in proportion to the length of the incoming airwave, something the kite bobbing in the breeze could never manage. Upon arriving in Newfoundland, Marconi first tried using blimps, figuring that they would do a better job of hovering in place, but the fierce winter wind ripped them all away. That left the young inventor with no choice but to try his last and least likely option, the kite-based plan that even one of his most ardent admirers labeled "not so much bold as hare-brained."

When setting previous distance records, Marconi had often

brought along pairs of skeptical journalists, allowing them to send prearranged codes to verify the airwaves were really working as Marconi claimed. This time, however, he had with him a "telegraph inker" that would sense three dots being tapped out on a telegraph key in England, then mark them on a strip of telegraph tape. If it worked, he would have some physical evidence for his claims. If it didn't, he would be spared the public humiliation of a reporter witnessing his failure. As the kite bobbed in the air outside, the inker refused to cooperate, forcing the inventor to try his last, desperate hope: a pair of telegrapher's headphones designed to turn incoming airwaves into audible dots and dashes.

After a few hours of work, Marconi and his longtime assistant, George Kemp, emerged from the small hut with remarkable news: while he had no physical evidence, Marconi declared that he had heard the three dots representing the letter S at 12:30, 1:10, and 2:20 P.M., a claim supported by his assistant. With that, Marconi and Kemp packed up their kite, leaving town without waiting to repeat their experiment in front of an outside observer.

Marconi's claim to have sent a single letter across the Atlantic hit the governor's office and quickly spread worldwide, flashing to the king of England and then onward to the global press, carried by the telegraph cables he aimed to replace. The *New York Times, The Times* (London), and *Le Monde* all splashed the news across their front pages.

While much of the press treated the claim as genuine, a backlash instantly followed. Many scientists, including Thomas Edison, openly scoffed at the claim, labeling Marconi a fraud or a fool. Amid the hail of static coming through a wireless headset, three little clicks would be difficult to isolate but easy to imagine. "I told *The Herald* last night that I doubted this story and I haven't changed my opinion," said Edison. "I don't believe it."

Most of the fiercest skeptics came from the Italian's home continent. By 1901, scientists had seen waves travel past the horizon, making a transatlantic transmission seem feasible. Such a feat is "simply a question of time and ingenuity" declared one French scientist in *Le Figaro,* before going on to call Marconi's take-my-word-for-it evidence laughable and his excuses for not being able to duplicate the feat far too convenient: "I, in common with all experts, refuse to believe that the single experiment adduced by Mr. Marconi was carried out under such precise, rigorous and unimpeachable conditions as to justify one in awarding the prize for its solution."

Even at his namesake company, the news of the founder's unexpected public announcement generated a concern that bordered on panic. When a *Brooklyn Eagle* reporter surprised the lawyer for American Marconi with the news of Marconi's claim, the attorney barely bothered to contain his disbelief. "What an ass that man Marconi is. He talks too damn much," exclaimed the lawyer. Even the flustered lawyer's efforts at more supportive sound bites reeked of doubt: "If Marconi says it is, it is. But I'm the senior counsel and it is not my place to say."

In his effort to convince the world of his claim, Marconi realized he possessed a critical advantage: most people wanted it to be true. In an era before antitrust laws, the transatlantic cable companies operated as an open cartel. Prices had been fixed at twenty-five cents per word since 1886, even though that meant 75 percent of the cables' information-moving capacity went unused. "For years there has been no advance in the utility of the cables, as measured by lower rates," complained a

New York Times editorial that captured the public disdain for the cable companies. "To all appearances they were as capable of improvement as the Martian canals and were managed with about as much reference to the needs and wishes of the population of earth."

At a dinner thrown in Marconi's honor by many leading American scientists, the decorations had included an elaborately carved block of ice, topped by miniature telegraph poles and broken wires. (The conceit being that soon the widely despised undersea-telegraph cartel would be "frozen out" by Marconi's new device.) Marconi moved quickly to capitalize on the implicit endorsement. The Marconi Company took out newspaper ads hyping the new subsidiary's stock. "Nearly 100 patents taken out by Marconi cover every phase of the electromagnetic wireless technology," the company bragged. Messages have been sent across the Atlantic Ocean, and even the great Thomas Edison endorsed the technology, the ads baldly declared. It worked: the Marconi Wireless Telegraph Company of America's initial public offering in 1902 raised $1.2 million in exchange for a slender 18 percent slice of the new firm, with its British parent company keeping the rest.

AS THE *LUSITANIA* headed up the Hudson River toward the Cunard docks on Manhattan's West Side, a *New York Times* reporter barraged Marconi with questions. The unscheduled interview gave the inventor a chance to practice the sales pitch he hoped would win over investors.

True, the company's transatlantic business had proven more difficult than he had expected a decade ago, Marconi conceded.

Now, however, its wireless technology had advanced to the point that the British government had agreed pay for the construction of a chain of wireless stations to connect the far reaches of its empire. Despite the United Kingdom's control of its own network of undersea telegraph cables—a network often referred to as "the nervous system of the British Empire"—the Crown was so convinced of the merits of worldwide wireless that it was paying to build an overlapping wireless network to move text messages around the globe. A decade of unmet promises made skepticism inevitable, Marconi allowed, but now even astute, practical investors could see that his promises of huge profits from transatlantic texting were bound to come true.

"So far we have constructed stations at Egypt and at Aden in Arabia at the mouth of the Red Sea, and also in Upper Egypt, which will communicate with Pretoria," Marconi boasted as the *Lusitania* steamed up the Hudson. "Later we are going to build stations in India, Singapore, Hong Kong, Australia, New Zealand and the West Coast of Africa. It is a very big undertaking."

The deal Marconi had struck with the British government was almost too good to be true—at least for shareholders in the British parent company. Marconi not only had beaten out rival wireless companies but also convinced the post office to pay £60,000 per station, with a lifetime of royalties to follow. The sky-high fees, so generous that they would soon become the subject of a parliamentary investigation, came after Marconi argued that the Crown could hardly trust its most sensitive communications to companies controlled by other countries. (Winston Churchill, the British home secretary at the time, bemoaned the hefty price tag but argued there was no better way to handle the "awkward situation" that results "when the question arises quite nakedly whether you should pay more to a British firm than you would pay to a foreign firm for exactly the same quality goods.") The British government's generosity was now reflected in the parent company's share price, which had quintupled thanks to the new contract.

While the British contract would not directly benefit the American Marconi Company, Marconi framed the deal as a chance for American investors to also get rich. All the U.S. subsidiary needed to do, he told the reporter, was raise the money to build wireless stations in the Caribbean, the East and West Coast of the United States, and Hawaii. Once the company linked those stations to the imperial chain, both arms of the Marconi Company would share the fortune that would come from replacing the worldwide web of wires with a worldwide web of waves—first in the main transatlantic route and soon along dozens of other lucrative routes throughout the western hemisphere.

Marconi felt so confident of his inevitable victory over the cables that he had begun to publicly speculate about making the system so cheap and powerful that it would allow him to supplant the main system for moving information across oceans: the mail. In recent interviews, he had discussed his dream of lowering prices to a penny per word (4 percent of the going cable rate) and, in the process, supplanting letters as the main conduit of transatlantic communication. Now, however, he restrained himself, focusing on the reasons he finally would vanquish the cable cartel.

STEAMING UP THE Hudson, the RMS *Lusitania* was an impressive sight. On her maiden voyage, hundreds of thousands of

New Yorkers had swarmed the banks of the Hudson River to gawk, paralyzing lower Manhattan with an epic horse-and-buggy traffic jam. Five years later, the news of the *Lusitania*'s arrival still made the newspapers, and the sight of her still could stop traffic. Today, as the afternoon sun sank low in the sky, the overdue ocean liner finally came in view of her home at Pier 54. She began a hard turn, four giant propellers churning the river, slowly swinging the ship into her familiar home.

The *Lusitania* displayed surprising nimbleness as she came about, a move made possible by her specially designed keel, which the British Admiralty had insisted upon in case the giant ship was one day pressed into military service and needed to dodge torpedoes. As the ship edged closer to the pier, a small fleet of tugboats stood ready to nudge her into berth while crewman lugged giant hawsers to secure the ship into place. Except for a few broken porthole windows, cracked in the mid-Atlantic storm that had sent huge waves smashing over the ship's bow, there was little evidence of her storm-tossed voyage. At half past three the gangways were thrown out and passengers began streaming off the ship.

Stylishly dressed as always, Guglielmo Marconi sauntered down the gangway into the first-class arrival room. Now age twenty-seven, the Italian inventor cut a dashing figure. His chiseled features, steel-blue eyes, and impeccable fashion sense made a particularly strong impact on New York society. ("[I] did not in the slightest resemble the popular type associated with an inventor in those days in America, that is to say a rather wild-haired and eccentrically costumed person," Marconi noted during an earlier visit.) Many people addressed him as Signor Marconi, a nod to his Italian homeland. To others he passed effortlessly as an English gentleman. Thanks to his Irish mother and Italian father, Marconi could transform himself from "Guglielmo" in Bologna to "William" in London or New York.

Two men accompanied him. The first, a narrow-shouldered gent with unusually large ears and a shock of dark hair rising above a high forehead, was Godfrey Isaacs, newly appointed to run the British Marconi Company and free the company's founder from the details of day-to-day management. The other man at Marconi's side, Percy Francis Heybourn, was a British "stock jobber" from the firm of Heybourn & Croft. In the London stock market of 1912, the main role of jobbers like Heybourn was to serve as the middlemen who brought together the buyers and sellers of a stock. In an era before dedicated "investment bankers," jobbers also advised companies on issuing new stock, which was why Heybourn had accompanied Marconi and Isaacs on this trip.

While Marconi's talent as an inventor—or lack of talent—remained the subject of bitter debate, his talent for hiring people with skills he himself lacked no one could dispute. At the start of his career, Marconi's key hires had been engineers like John Ambrose Fleming, whose work had multiplied the abilities of the Marconi Company's early gear. Godfrey Isaacs, had proven another excellent hire. A lawyer by training, Isaacs had developed an obsession with controlling as many patents as possible. Already, various inventors had won dozens of patents covering the many parts of a wireless communications system, including critical pieces of antennas, transmitters, batteries, receivers, and more. Before hiring Isaacs, Guglielmo Marconi had made a few scattered patent purchases himself, using his investors' cash to buy them from Thomas Edison, among others. Isaacs considered patent buying his primary mission. His greatest coup had been persuading two of Marconi's fiercest academic critics, Professors Oliver Lodge and Alexander Muirhead, to sell their crucial patents to the Marconi Company. The Lodge-Muirhead patents included a key breakthrough that allowed a receiver to isolate or "tune into" airwaves of a specific length. That made it possible for different receivers to pick up different wavelengths and the airwaves to move multiple messages at the same time.

Lodge and Muirhead, like many members of the British professoriate, viewed Guglielmo Marconi as a charlatan who had patented a compilation of other people's discoveries. Isaacs won them over all the same. Tough but soft-spoken, he made it clear that they could either to sell to the Marconi Company (and become rich immediately) or hire a lawyer (and take their chances on a long, expensive legal fight). Isaacs crafted an offer sweet enough for the professors to set aside principle. Professor Lodge cashed a check for £18,000 (roughly £1 million in 2015) and joined British Marconi Company as a well-paid consultant.

Guglielmo Marconi's visits to New York invariably revolved around three things: money, the airwaves, and women. During past visits, the company's messenger boy, David Sarnoff, had handled the delicate tasks of delivering the right flowers, candy, letters, and other romantic gestures to the correct recipients. (Marconi behaved, in the words of one contemporary, like "a Byronic hero transplanted to 20th century New York.") The famous inventor appreciated the tact and savvy with which Sarnoff managed his active and overlapping social life. His young employee's eager curiosity reminded him of himself as a teenager. Every technical and financial facet of the business seemed to interest the former messenger boy, who used his informal role as Marconi's American assistant to barrage his boss with questions during any moment of downtime.

On this trip, however, Marconi expected to have little time for anything but selling his new plans to wary investors.

To bolster the reputation of his transatlantic service, Marconi arranged another demonstration for the press, though it too would leave room for doubt and soon cause its own controversy. He arranged for a flurry of famous names (who happened to all be big fans of his) to send a series of "Marconigrams" from London to a dinner party in New York. The *New York Times* reported a new speed record set by the message Lord Blythe sent just after midnight in London that arrived at the dinner ten minutes later, 7:25 P.M. New York time. The message: "Of all the wonderful discoveries the world has ever seen, none is to be compared with the fabulous invention of wireless telegraphy. All honor, therefore, is due Marconi, to whom we are under a lifelong debt of gratitude."

To skeptics, the stunt actually demonstrated why the Marconi Company's competitive problems went far beyond the basic ability to send text messages over the ocean. The Marconi Company couldn't offer Wall Street traders anything approaching a ten-minute delivery time during daylight and business hours. For the dinner, Marconi had made a special request of the British Post Office, which ran that country's telegraph system, to expedite the messages' delivery from London to Marconi's wireless transmitter in the south of England. (The previous London–to–New York record was fifty-five minutes.) And by now it was well-known that the ability to send a wireless message at night, when that the signal-killing sun was on the other side of the world, did not guarantee the same feat would be possible during business hours.

THE DAY BEFORE American Marconi shareholders were scheduled to vote on Guglielmo Marconi's proposal, a knock at the door interrupted the inventor's dinner. A minute later, hat and coat in hand, he sprinted out the door. The *New York Times* reporter who had interrupted his meal suggested they take the Ninth Avenue elevated train, which offered the quickest way down to the dock that had welcomed Marconi and the *Lusitania* a few weeks earlier.

As their train reached Fourteenth Street, the smell of the meat-packing district's slaughterhouses filled the air. Continuing on foot, Marconi and the reporter hustled west, toward the Hudson River. Soon Marconi began to see groups of distraught friends and relatives, clustered beneath signs with large letters. At the sight of a young lady sitting on an old woman's lap, both of them crying as the older woman kissed the younger one's forehead and stroked her hair, Marconi paused for a moment. The *Times* reporter noted tears in the inventor's eyes as he climbed aboard the ship that now occupied the *Lusitania*'s berth, and headed for the upper deck, where the steady *pop-pop-pop* of a spark-gap transmitter at work led him to the Marconi Room.

Inside, illuminated by the glow of a single light, he saw one of his employees, Harold Bride, a gaunt figure with heavily bandaged feet, tapping away on his telegraph key.

"Hardly worth sending now, boy," Marconi said. The junior operator, whom Marconi had never met, seemed to recognize his boss from the picture hanging on the wall. The men shook hands in silence. Then Bride began to talk quietly, telling the story of the tragedy that would turn Guglielmo Marconi into a national hero and guarantee the financial future of his company.

CHAPTER 4

BRIDE

The North Atlantic—1912

HAROLD BRIDE AWOKE JUST AFTER MIDNIGHT FOR HIS OVERnight shift, sleepy but otherwise untroubled. He had been expecting a busy night. Earlier that day the ship's wireless equipment had gone out and he and his boss, senior operator Jack Phillips, still hadn't yet worked through the backlog of messages that had built up in the hours they had spent diagnosing and fixing the problem. At that moment, though neither he nor Jack Phillips had realized it, the two Marconi operators offered the only hope of survival for the 2,223 people sailing on the maiden voyage of the RMS *Titanic*. The iceberg that had sliced through the ship's hull cut so cleanly that the junior Marconi operator had slept right through it.

The wireless technology on board the *Titanic* was, at a basic scientific level, exactly the same as the gear Heinrich Hertz had used to conjure up airwaves a quarter century earlier: sparks jumping between two pieces of metal. As a practical matter, however, spark-gap technology had come a long way, and in building the *Titanic,* no expense had been spared. Twenty-five years of engineering tweaks to Hertz's rudimentary transmitter had made an art of summoning artificial lightning bolts. The effects were easy to see, and hear. The modern Marconi system drew five kilowatts of power straight from the *Titanic*'s main generator, which it then fed into a spinning metal disk. Thanks to a clever system of studs on the spinning disk, it was possible to create split-second sparks, allowing skilled fists to tap out messages quickly in Morse code.

The *Titanic*'s receiving equipment also was state-of-the-art. High above the main deck, four wires—the antenna—made from an alloy of silicon and bronze ran nearly the entire length of the ship, suspended above the *Titanic*'s four huge funnels by a pair of even taller masts. The 120-meter antenna was designed to be precisely one quarter the length of the invisible waves that Marconi transmitters used, a proportion that allowed the antenna to more easily resonate as the incoming electromagnetic vibrations struck it. The wires ran down to the roof of the Marconi room. From there another wire fed down through a pillar and into the room where Bride and Phillips worked, delivering the invisible waves to the receiving set that Phillips held to his ear.

Neither man suspected the extent of the crisis until Captain Edward Smith appeared at the door, a blue hat covering his full head of gray hair. "We've struck an iceberg," the captain announced, without preamble.

He had ordered an inspection, Captain Smith continued. "Get ready to send out a call for assistance, but don't send it until I tell you to."

When the captain headed back the bridge, Phillips got back on the receiver, searching for any ships in the vicinity. Ten minutes later, the captain reappeared, barely sticking his head in the door.

"Send the call for assistance," he ordered.

"What call should I use?" Phillips replied, hoping that perhaps the captain wasn't referring to the call reserved for the gravest of emergencies.

"The regulation international call for help," the captain snapped, before turning on his heel and heading back to the bridge.

Dash-dot-dash-dot, dash-dash-dot-dash, dash-dash-dot . . .

As the junior operator, Harold Bride stood by while Jack Phillips worked the key, reflexively translating the make and break of the telegraph key into English letters.

CQD. The standard distress call.

DE. Fists' shorthand for "this is."

MGY. The *Titanic*'s call sign.

Position 41''46N, 50'14W.

The only vessel with comparable wireless range was the *Titanic*'s sister ship, the *Olympic*. The Marconi Company published a schedule of ships crossing the Atlantic, but a quick check of the *Olympic*'s location gave Bride and Phillips little hope that she could arrive in time to help.

In fact, thanks to her own giant antenna, the *Olympic* had caught the shocking message, though all her horrified Marconi operators could do was relay the news to a sleeping world:

```
Titanic sending out signals of distress.
Answered his calls.
```

Titanic replies and gives me his position 41.46N 50.14W and says "We have struck an iceberg."

Our distance from Titanic: 505 miles.

Just as Bride and Phillips were digesting this bitter news, they caught a break. Harold Cottam, the sole Marconi operator aboard the RMS *Carpathia,* had missed the *Titanic*'s first distress call while delivering a message to the bridge. He had, however, overheard several messages intended for the *Titanic* from the Marconi station in Cape Cod earlier in the day. Before heading to bed, he decided to see if his colleagues wanted him to forward them along.

Do you know Cape Cod is sending a batch of messages for you?

Phillips cut him off instantly:

Come at once we have struck a berg.

It's a CQD OM. [It's an emergency, old man.]

Position 41.46N 50.14W.

Cottam—wide-awake now—replied:

Shall I tell my captain? Do you require assistance?

Yes come quick.

Using the *Carpathia*'s coordinates, as Cottam had just relayed them, Phillips and Bride calculated that the midsized liner was just fifty-eight miles (ninety-three kilometers) away. The much smaller ship had been traveling the less popular route from New York to Austria-Hungary, but its path toward the Strait of Gibraltar had left it tantalizingly close by.

Among the *Titanic*'s many modern touches was a telephone system with fifty lines, one of which connected to a handset on the table near where Bride was standing. No one, however, had thought to install a direct line to the bridge. So while Philips worked the telegraph key in an attempt to find other nearby ships, Bride, still in his pajamas, went to find the captain, dashing down a corridor that ran past the officer's quarters on the ship's port side. Though Captain Smith had said the *Titanic* had been struck amidships, Bride could feel the ship tilting downward, bow first. He arrived on the bridge to deliver the news of the inbound *Carpathia* to the captain.

Returning to the Marconi room, Bride threw on warmer work clothes and an overcoat. He also brought a coat to Phillips, who hadn't left his station and had continued furiously tapping out updates to the *Carpathia* as the *Titanic*'s forward list became more pronounced. The captain stopped in to deliver another update: the ship's generators were being swamped. Phillips relayed the information to the inbound *Carpathia*, now charging through the seas toward them, her men frantically shoveling coal into the burners, pushing the older ocean liner to her limit.

A few minutes later, the *Titanic*'s main power died. Captain Smith returned a final time to release Bride and Phillips from service, telling them there was nothing more they could do. Then he disappeared. Neither Marconi operator was ready to give up. While Phillips tried to coax a last string of sparks from the wireless, Bride found the two men's life preservers, attaching one to Phillips as he worked. Bride could tell Phillips was straining to hear faint incoming dots and dashes. This time, the reply was too quiet for Bride to overhear.

By the time the last of the emergency power gave out, the ship had pitched forward at a terrifying angle. Bride and Phillips struggled to make their way outside. Nearby, on the A-deck, a handful of passengers were wrestling loose a final collapsible lifeboat, trying to get it off the deck. Bride stayed to help out. Phillips headed aft to look for better options. Not long after, Bride and his fellow passengers managed to knock the collapsible boat into the ocean, Bride plunging in with it. He landed in the frigid water directly underneath the overturned lifeboat.

Fighting his way to the surface, he swam away from the sinking ocean liner, terrified of being sucked down with it. Instead, the *Titanic* slipped beneath the water with an ease that reminded him of a duck going down for a dive. With the *Titanic* gone, Bride swam slowly back to the overturned lifeboat, where another passenger helped him clamber aboard. Even with the numbing cold, Bride could feel pain shooting up his legs.

A high-pressure zone kept the winds light, the sky clear, and the sea glassy. The young Marconi operator passed on word of the group's only cause for hope. The *Carpathia,* he calculated, was by now less than thirty miles away. Someone aboard the overturned lifeboat suggested a prayer, and after a survey of the group's religions—"Presbyterian," "Catholic," "Methodist"—they settled on the Lord's Prayer as the ecumenical choice.

Bride scanned the water for Jack Phillips. The senior Marconi officer was nowhere to be seen.

Of the *Titanic*'s 2,223 passengers, 1,517 never made it to shore alive. Thanks to the Marconi operators and their wireless telegraph, 706 people did.

AS SOON AS word of the Marconi operators' role in summoning the *Carpathia* reached New York, the grief-stricken city embraced Guglielmo Marconi as a hero. His long-scheduled technical lecture in the Engineering Societies Building had been overwhelmed by an overflow crowd, who welcomed him with a two-minute standing ovation. Even the once-skeptical Thomas Edison had come around. This time there could be no doubt of his intent in the letter he wrote to Marconi "to congratulate you upon the success of your beautiful invention—the wireless telegraph—and on the splendid work your system has done in saving human life in disasters on the sea."

The morning after his visit with Harold Bride aboard the *Carpathia,* Marconi headed to the annual shareholders meeting of the Marconi Wireless Telegraph Company of America. The shareholders' mood was both somber and celebratory. The vote that only a week ago had seemed so uncertain—whether to risk another $7 million in newly issued stock to build new stations designed for intercontinental telegraphs—was now a fait accompli. That the investment would have little to do with saving lives at sea seemed to scarcely matter. The power and value of the airwaves had been seared into the minds of the nation.

Most shareholders just seemed happy to applaud whatever pronouncements the heroic inventor wanted to make.

"If Guglielmo Marconi were not one of the most modest of men, as well as of great men," declared that morning's *New York Times,* reflecting the mood of the city, "we would have heard something, possibly much, from him as to the emotions he must have felt when he went down to the Cunard wharf, Thursday night, and saw coming off the *Carpathia*, hundred after hundred, the survivors of the *Titanic*, every one of whom owed life itself to his knowledge as a scientist and his genius as an inventor."

When all the votes were counted, nearly 75 percent of shareholders voted to put up another $7 million to fund another generation of wireless stations in the United States.

Just as he had a decade earlier, Guglielmo Marconi sailed out of New York Harbor in triumph: his bank account stuffed with dollars, his head full of grand visions. Whatever his limitations as an inventor, as a businessman Marconi justly could claim credit for discovering one critical feature of the wireless communications industry. It wasn't exceptionally talented inventors who tended to wind up in the control of the invisible waves, but rather exceptionally talented liars.

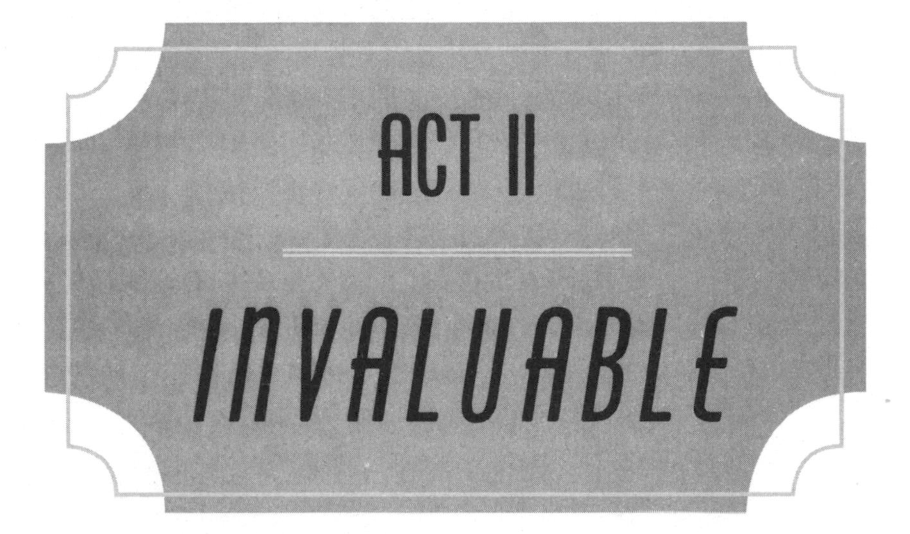

ACT II

INVALUABLE

CHAPTER 5

SARNOFF

New Jersey—1914

DAVID SARNOFF TURNED OFF THE MAIN ROAD, HEADING TOward the New Jersey coast. Nearing the town of Belmar, where the Shark River pushes inland, he finally spied what he had been looking for: the top of a three-hundred-foot-tall silver tower peeking into view above the tree line. Seconds later, an identical tower appeared above the treetops. Soon a row of shimmering steel towers running for more than a mile came into view, holding aloft a long, thin wire. It was the largest radio antenna ever built—the world's largest electromagnetic ear—and David Sarnoff could hardly wait to find out what he would be able to hear when he hooked it up to Edwin Howard Armstrong's new invention.

Sarnoff had several reasons to be in a sunny mood. A few months after the *Titanic* sank, when he gave up his job managing American Marconi's flagship office to take one in the engineering department, many of his colleagues figured that his meteoric rise had stalled. His new job as an inspector with American Marconi's engineering department turned him into a glorified repairman, responsible for making sure wireless telegraph transmitters and receivers were in good working order when the ships carrying them left port. What others saw as a step down, Sarnoff considered a chance to see how all the wireless gear actually worked, in the real world. Inspectors got to travel the eastern seaboard, seeing any new gear in action and debriefing Marconi operators when they hit port. The wisdom of that career decision had been borne out by his recent promotion, to chief inspector, a bump-up that put the twenty-two-year-old in charge of a small army of older engineers. These days, the employees manning the wireless telegraph keys didn't dare belittle him or his religion, not to his face, at least. The days of being called "Davey" and "Jew Boy" were behind him. It was "Mr. Sarnoff" now.

While David Sarnoff's ascent had continued at the Marconi Company, so had his famous boss's penchant for sending the company careening toward bankruptcy. The reason was still the same. Thirteen years after he first claimed success on a Canadian hilltop, Guglielmo Marconi still struggled to transmit signals across the Atlantic Ocean. Like his mentor, Sarnoff dreamed of the day when wireless signals could match the cables' speed, capacity, reliability, and profitability. Unlike Marconi, the young chief inspector had the practical experience and emotional distance to realize that despite the spending spree of the last two years, the company was not nearly there.

Today, David Sarnoff had cooked up an ambitious plan to prove his value to Guglielmo Marconi. If he was right, Edwin Armstrong, the tall, gray-eyed inventor accompanying him, held the key to rescuing the Marconi Company in his pocket.

Taken as a group, Sarnoff, Armstrong, and the two men walking with them below the giant antennas possessed an unparalleled knowledge of the airwaves and the tools to control them. Roy Weagant, a thirty-two-year-old Marconi Company colleague, had been Sarnoff's main rival for the chief inspector job. The older engineer had made several clever inventions himself, and though none of them had proven good enough to make transatlantic texting workable they sealed Weagant's reputation as a top-flight technical mind.

If his boss's decision on whom to promote had come down to pure technical talent, Sarnoff would have lost to Weagant. He won the chief inspector job thanks to his combination of technical knowledge, business savvy, and people skills. His continued friendly relationship with Weagant, including his decision to invite him along on the evening's adventure, offered the latest example of David Sarnoff's knack for winning over allies, even an older man he had just beaten out for a promotion.

Near Weagant sat another thirty-two-year-old. This one, Sarnoff knew less well. Professor John Morecroft, from Columbia University, had a reputation as the star protégé of academia's most famous and talented electrical engineer, Michael Pupin. These days, even Professor Pupin looked to Morecroft for guidance when it came to building and using the latest devices for summoning and detecting invisible waves in the air. Of the four men traveling to Belmar, Morecroft had the greatest knowledge of the theoretical physics that described how electromagnetic vibrations move through space. Among his Ivy League colleagues, however, Morecroft was best known for his practical know-how. Before becoming an academic, he had worked as the foreman of a machine shop. That, no doubt, was why his former student Edwin Armstrong had chosen to invite him as his guest for the evening's excitement.

Of his three companions, it was the twenty-three-year-old who most intrigued the chief inspector. At first blush, Edwin Howard Armstrong and David Sarnoff had little in common beyond their age. Sarnoff was a Russian immigrant born to impoverished Jews; Armstrong grew up in the New York suburbs, the scion of well-to-do Presbyterians. Sarnoff was short, Armstrong tall. Sarnoff radiated a self-confident charm, Armstrong an introvert's awkwardness. Sarnoff's formal education ended in eighth grade; Armstrong had just graduated from Columbia University with a degree in electrical engineering. And yet, when the two talked about the power and possibility of the invisible waves that obsessed them, those superficial differences melted away. Both men had loved tinkering with wireless gear since they were boys. Both put little stock in abstract scientific theories, believing that practical experiments were far more likely to reveal new discoveries.

Professor Morecroft had watched Armstrong earn his Ivy League degree almost in spite of himself. As a student, Armstrong had risked expulsion several times by embarrassing Morecroft's colleagues, finding sly ways to use his practical knowledge to publicly expose the professors' mistaken theories. One famous story ended with his goading a professor into touching a wire that, according to the teacher's theory, was harmless. That professor left the classroom with singed fingertips and a lesson in trusting theory over the word of Edwin Armstrong.

Much like Sarnoff's boss and idol, Guglielmo Marconi, Armstrong seemed refreshingly uninterested in Sarnoff's religion or social class. To Armstrong, David Sarnoff was that rare bird whose enthusiasm for the potential of wireless technology matched his own—and who possessed an even greater familiarity with the tools that had been invented to control the airwaves.

BEFORE GETTING DOWN to work, Sarnoff and Weagant gave Armstrong and Morecroft a quick tour around their company's facility. The two-hundred-acre campus was unlike anything the two men had ever seen. Now visible in their entirety, the mile-long row of three-hundred-foot-tall steel masts cut across the property in a straight line, holding aloft a barely visible wire. Custom-engineered to withstand gale-force winds, the towers were a world away from the shabby collection of blimps, kites, and rickety wooden masts that the Marconi Company had relied on to elevate its antennae a dozen years earlier.

Guglielmo Marconi had been spending his investors' money lavishly in other ways as well. The property's next-largest structure, a new red-brick hotel, could house up to forty unmarried telegraph operators, who would work in shifts encoding and decoding the transatlantic traffic. Extravagant touches seemed to be everywhere. Operators with families would live in one of the four-bedroom cottages that dotted the grounds. Ornamental gardens, designed by a landscape artist, offered views of the river. The sprawling property also had places for off-duty staff to hunt and fish. After a long shift, the operators would be able to repair to the hotel dining room, where a French chef would cook their meals.

Neither Armstrong nor Morecroft had any professional relationship with the Marconi Wireless Telegraph Company of America or its British parent company, but both men followed its progress closely, naturally curious for any news of the world's largest wireless company and its famous founder. Since the sinking of the *Titanic,* most of that news had centered around Marconi's push to build the Imperial Wireless Chain and his American subsidiary's construction of a complementary (though nonimperial) chain of wireless stations in the western hemisphere. As the tour made plain, Marconi was spending his shareholders' money assuming that he would not just compete with the transatlantic cable cartel—he would conquer it.

In front of his guests, Sarnoff had no reason to dwell on his own concerns about Marconi's construction spree. Instead he steered the discussion toward the many genuine innovations the Marconi station had to show off to its guests. Most obvious were the tubular steel masts, which cut across the property at an odd, seemingly arbitrary direction. The British Marconi Company, Sarnoff explained, was putting the finishing touches on a sister station on a mountaintop in Wales that was being kitted out with the newest, most powerful transmitter available, capable of blasting out waves with four hundred thousand watts of power. The wire overhead had been aligned to run perpendicular to the direction of incoming Welsh waves, the better to absorb every jot of power remaining after the long trip across the Atlantic.

A quarter century after Hertz invented his spark-gap transmitter, various inventors around the world had begun to discover new ways to generate the invisible waves—building a second generation of mechanical mouths able to speak the voice of lightning with far greater fluency and at far greater volume. The first-generation spark-gap transmitters suffered from several major problems, the details of which Sarnoff didn't need to elaborate for his technically sophisticated guests. Each spark in the air released a burst of energy, an electromagnetic cacophony of varying wavelengths analogous (in the world of sound waves) to the noise made by a flat hand slammed down onto piano keys. As a result, the spark gap transmitters spoke a language made up of a single monosyllabic grunt. The only way to use that language to convey information was by varying the grunts' duration. A brief spark created a brief grunt that a distant receiver could hear as a Morse code dot. A spark that hung in the air a split second longer created a grunt that also lasted a bit longer, creating a dash.

The new generation of transmitters, by contrast, gave the voice of lightning a new vocabulary. Instead of sparks in the air and indiscriminate blasts of electromagnetic energy, the new machines used a high-speed electric motor to generate a continuous electromagnetic wave with a single, fixed length: usually several kilometers from crest to crest. To the four experts touring the Marconi station, the potential of these "continuous wave generators" was thrilling. By focusing power into a single wavelength (instead of dispersing it across many different waves of many different lengths), the generators could send messages much farther. And, in addition to carrying Morse code, the waves could now be modified to carry sounds, just like a telephone.

From a business standpoint, the continuous wave generators opened up the possibility of transmitting lots of waves with different lengths simultaneously, each of them ferrying a different message. Thousands of different operators might then be able to "tune into" the waves, radically increasing the airwaves' power to connect different places around the world.

To Armstrong and Morecroft's disappointment, the new

transmitters the company was installing were nowhere to be seen. A transmitter here on the Jersey Shore, shouting eastward, would drown out the westbound whispers the Belmar station was straining to hear. So, to avoid crossed signals, the company had positioned its American transmitter on the other side of New Jersey. Similarly, the European sending and receiving stations sat on opposite sides of Wales. The Marconi operators in Belmar, New Jersey, and Caernarfon, Wales, were able to control the distant transmitters remotely, using a land-based telegraph line that connected the sending and receiving stations.

In numerous press accounts, Marconi had declared his complete confidence that once the stations in New Jersey and Wales were complete, all his long-standing challenges finally would be licked. The lavish, state-of-the-art station in Belmar and its sisters had been designed to solve every problem that had dogged his twelve-year effort to send transatlantic messages. Sending wireless telegrams between Wales and New Jersey, instead of Cornwall to Canada, would circumvent a longtime bottleneck. The company's text messages would no longer need to be relayed from Newfoundland to New York over telegraph cables owned by the companies Marconi was vowing to destroy. Service during business hours also was coming, Marconi promised, thanks to his new station's transmitters blasting out waves so strong that they would overpower daylight's signal-deadening effects on the ionosphere. Soon a British stock trader's order would whiz across the ocean at the speed of light, hit the antenna in Belmar, get transcribed by a Marconi operator, and then immediately be forwarded to Wall Street over a special, Marconi-owned telegraph cable.

WITH NOTHING ELSE TO SHOW OFF, Sarnoff and Weagant brought their tour of the Belmar station to a close. Interested as the four men were in the new station and its innovations, they were far more excited about the experiment they had come together to conduct.

The lanky Armstrong, while cagey about the exact details of his device, was very clear that it wasn't merely one more incremental improvement. Instead, it was a new sort of device altogether—a revolutionary tool capable of feats the scientific community considered impossible. Armstrong claimed that his invention was able to grab hold of an invisible airwave and alter one of its fundamental characteristics: its amplitude. The greater the distance from the crest of a wave to its trough, the greater the power it carries. Waves in the ocean offer an easy way to visualize the relationship of height to power. A ten-meter-tall wave crashes onto the beach with a thunderous boom; a tiny ripple gently laps against the shore. Amplitude and power have a similar relationship in sound waves (ears recognize tall sound waves as loud noises) and electromagnetic waves (eyes recognize "tall" light waves as bright light). All four men were also extremely familiar with another universal property of waves: they expend energy as they move along. Traveling along the surface of the ocean, even the tallest waves gradually lose power and shrink to a ripple.

Before Armstrong, inventors hoping to use the airwaves to communicate across ever-longer distances had few options to offset the problem of waves' fading power. The most obvious approach was simply to start with a bigger wave. Crossing greater distances with ever-more-powerful electromagnetic waves had always been Marconi's preferred strategy. (Sound waves again offer a familiar analogy: if speaking in a normal voice isn't audible to someone in the distance, shouting creates a taller sound wave that travels farther.) The second solution, nearly as obvious: build a bigger ear. Larger antennae help by collecting more of the wave's faded energy. The mile-long antenna on the Jersey Shore was essentially a giant metal ear cupped to capture faded messages shouted from across the sea.

Radio engineers' final trick was to improve the "eardrums" used to sense incoming electromagnetic waves, developing more sensitive instruments able to pick up fainter signals. A famous example of this tactic sprang from the mind of Lord Kelvin when he was trying to get the first telegram through a cable running beneath the Atlantic. To move a magnet with the miniscule amount of energy that survived the undersea trip, he suspended a tiny magnet from a silk thread and glued a small mirror to it. That way, even when the human eye couldn't see electricity make his micro-magnet twist, a beam of light focused on the mirror would whip weightlessly from side to side. A louder mouth, a bigger ear, a keener eardrum: to date that was how electromagnetic communications had improved.

Now Armstrong unveiled some tiny circuitry hidden inside a small wood box. It had, the inventor claimed, the power to take a faded wave streaming in from far away and reinfuse it with enough power to make it easily detectable.

A few weeks earlier, Sarnoff had witnessed a tantalizing trial of Armstrong's new device at Columbia University. Using a small antenna he had strung up between a few university buildings, Armstrong had briefly tuned in signals that seemed to originate in far-off San Francisco but then disappeared in a hail of static. Sarnoff left that demonstration intrigued but skeptical. Armstrong's initial test used a "particularly amateurish" antenna,

Sarnoff had told his Marconi colleagues. He also worried that while Armstrong seemed honest, it was always possible that he had stashed a confederate with a small transmitter on the other side of campus to fake the San Francisco signals. Nonetheless, that test had started Sarnoff thinking about the magical combination that might result from hooking up Armstrong's new invention to Marconi's world-class antenna in Belmar.

At 4 P.M., with the afternoon sunlight beginning to fade, Sarnoff led the group into the unheated shack where the antenna wires terminated. He decided to start off with an easy test. Using the station's standard equipment, which wasn't connected to Armstrong's invention, he tuned into the company's station in Glace Bay, Canada. The signal came in clearly, so clearly that it could be heard even when the telephone earpiece sat on the table. Next, they attached Armstrong's device, which the inventor kept concealed in a box, hooking it to the wires running up to the giant Marconi antenna.

Suddenly the incoming signals became not merely audible but deafening. Sarnoff and Weagant backed away—even at fifty feet they could easily make out the incoming message blasting through the headphones. Encouraged, Sarnoff turned to a harder challenge. The Welsh station wasn't transmitting yet, so he used the station's normal gear to tune in the older Marconi station in Clifden Bay, Ireland. Using the full mile-long antenna, Sarnoff was able to tune in weak, barely detectable signals. Plugging into the second, shorter antenna wire that the towers held aloft, which ran only 1,600 feet, made things worse, yielding a flurry of dots and dashes that faded in and out through a wall of static. Even with Sarnoff's expert ear pressed against the earpiece, many of the messages were indecipherable. Then Armstrong hooked up his mystery device. Immediately the Irish signal blasted from the headset, clear and strong.

Rather than hand the headphones to Sarnoff, Armstrong set them down on the table, looking on happily as Sarnoff and Weagant realized that the signal from Ireland was suddenly so powerful that they didn't need the headphones to make out the incoming message. The improvement was so mind-blowing that, as Sarnoff and Weagant jotted down notes, they struggled for a way to quantify it. The signals beaming in all the way from Ireland were, thanks to Armstrong's amplifier, several times as powerful as the signals the regular Marconi gear was picking up from neighboring Canada.

Around 8 P.M., as the signal-killing sun set on the far side of the United States, Sarnoff shifted the experiment's focus westward. The Marconi Company's stations on the other side of the country were still under construction, including a gargantuan antenna in Hawaii that would be even larger than the one here in New Jersey. So Sarnoff began searching for signals from the Marconi Company's Danish competitor, the Poulsen Wireless Telephone & Telegraph Company, which ran stations in San Francisco, Portland, and Hawaii. Neither Sarnoff nor Weagant could raise the Poulson station in California using Marconi's standard gear. The Frisco station's signal used a a mere nine thousand watts of power, about 2 percent of the power of Marconi's station in Wales. Giving up, they turned control over to Armstrong to try out his gizmo—and suddenly Sarnoff could make out dots and dashes traveling from San Francisco to Portland with such astounding clarity that he was able to copy down forty words a minute with ease.

So it went for the rest of the night, the men too excited to sleep, all of them grappling with the implications of the stunning leap forward. At one point, searching for a new signal, Sarnoff heard six dots followed by a dash, which he instantly recognized as the call sign of the Poulson station in Hawaii. He began to transcribe the conversation the Honolulu station was attempting to have with San Francisco, despite thunderstorms hitting the islands.

```
  Lightning bad. Shall ground aerial
wires.
  OK, will call you in 15 minutes.
```

The amazed Sarnoff realized that he was having an easier time picking up the messages sent from Hawaii than the San Francisco station that was trying to receive them, even though Sarnoff was sitting in New Jersey.

Occasionally, he and Weagant prodded Armstrong for more details of his unusual gadget. The young inventor deflected most of their queries.

At 1:25 in the morning, Sarnoff shifted focus back to the east, attempting to tune in the Telefunken station in Germany. Soon the telltale dots and dashes came through the headphones: POZ, the call sign for a transmitter near Berlin. Finally, at 5 A.M. the four men headed back to New York, giddy with the knowledge that the invisible world had revealed itself to be far more powerful—and far more valuable—than anyone outside their small group could possibly imagine.

To Sarnoff, Armstrong's invention was miraculous both in its technological achievement and in the timely chance for rescue it offered Guglielmo Marconi. For the first time, Sarnoff believed that the technology had advanced to the point that it could turn the airwaves into a legitimate threat to the undersea cable cartel. And Sarnoff, as the Marconi employee who had learned of the device and befriended its inventor, would be the one to provide his patent-loving bosses in London with the chance to buy it before the rest of the world recognized its remarkable power.

AFTER RETURNING TO New York, Sarnoff wrote a four-page memo to his direct boss, American Marconi's chief engineer. He knew his report would quickly make its way across the Atlantic and into the hands of both Godfrey Isaacs and Marconi himself. Displaying confidence in his ability to understand both the technology and its economic impact on the information-moving business, he focused on describing his tests of the amplifier while musing about how the Marconi Company could adapt and profit as a result.

While Sarnoff had built his career by displaying a maturity that belied his real age, his youthful enthusiasm peeked out. And why not? Given his expert ear, unparalleled knowledge of gear being used in the field, and careful double-checking, there was no doubt in his mind. A new revolution was at hand that would vastly increase the value of the invisible waves in the air.

"Unless there be other systems of equal merits, which are unknown to me," the Marconi Company's chief inspector concluded, covering his last base, "I am of the opinion that it is the most remarkable receiving system in existence."

CHAPTER 6

ARMSTRONG

London—1917

WHEN THE UNITED STATES ENTERED WORLD WAR I, PRESIDENT Woodrow Wilson ordered the navy to seize control of all American Marconi stations for the duration of the war. Not long after, the U.S. Army Signal Corps shipped its most prized recruit–the newly commissioned Captain Edwin Armstrong–to the front to see what he could do about the embarrassing state of military communications technology. American companies led the world in inventing and building gear to harness the airwaves. The country's armed forces, however, still relied on strategy and tactics designed around the limits of telegraph wires and carrier pigeons.

The war had made the airwaves' military importance clear from its opening day, when the British dredged up and cut the five telegraph cables connecting Germany to North America. German U-boats responded in kind, severing all the British cables. After cutting the cables, the British military took control of every Marconi Company wireless station in the Empire, along with the rest of the company's manufacturing operations.

With the cables cut, for the first time in the long and one-sided rivalry between waves in the air and wires under the sea, the airwaves had a distinct advantage. Besides connecting the war's newest combatant with its allies in Europe, the waves also linked battleships at sea and battalions of troops in the trenches, often providing critical tactical advantages over relying on telegraph and telephone wires (which could be cut) and fleets of carrier pigeons (which would only deliver a scrap of paper to a prearranged destination they had been taught to home in on). The U.S. Navy had just launched a crash program to get its transport ships out of the pigeon age. The army, widely acknowledged to be even further behind, created a new organization, the Division of Research and Inspection, to upgrade its technology, especially its wireless equipment. The army appointed Captain Armstrong to head the division's radio group.

CAPTAIN ARMSTRONG'S ORDERS directed him to Paris, where the Signal Corps had set up its headquarters. Along the way, he found himself temporarily stuck in wartime London with one evening to kill. To entertain himself, the new army officer decided to walk over to the city's theater district and see what was happening at British Marconi headquarters.

Located between the Royal Opera House and the river Thames, the Marconi House had long been a source of fascination to the American inventor. Guglielmo Marconi had purchased the seven-story building, previously home to one of London's largest restaurants, and converted it into a combination research center, manufacturing plant, and corporate headquarters. The renovation was part of the spending spree financed by Marconi's Imperial Wireless contract, but walking inside, Captain Armstrong could see none of Marconi's typically lavish touches. While the building's giant entrance hall hinted at its former life as a fancy eatery, the soldiers standing guard gave away Marconi House's new role as a British military station.

Despite his personal admiration for Guglielmo Marconi, an idol and role model since childhood, Captain Armstrong had mixed feelings about Marconi's various companies. Following his experiment with David Sarnoff at American Marconi's Belmar station, Armstrong had eagerly awaited the response from London to Sarnoff's enthusiastic report. The reply had stunned them both. Guglielmo Marconi refused to believe the detailed experiments, dismissing Sarnoff's report as "very incomplete" and mistakenly surmising that Armstrong's invention was very similar to technology Marconi had been developing. Godfrey Isaacs was less dismissive than irate, ordering the head of American Marconi to ban such unauthorized experiments for fear they would force up the price of outside inventors' patents.

Armstrong had been disappointed by Marconi and Isaacs's haughty dismissal, but the rejection had been even harder on his advocate inside the company. David Sarnoff idolized his boss and had expected to be praised for finding a device that finally could make Marconi's transatlantic dreams come true. Marconi's technological blindness shocked Sarnoff, with Godfrey Isaacs's public scolding adding a dose of humiliation. American Marconi president E. J. McNally made sure his entire company knew about Sarnoff's blunder, writing a letter to all

American Marconi engineers in which he declared himself to be "in absolute accord with Mr. Isaacs" about the perils of Sarnoff's unauthorized adventure, which "advertises the inventor at our expense, and creates a competition which means trouble and a menace."

Three years later, it was now much easier for Armstrong to laugh about Marconi and Isaacs's hubris. For a man who built his career on his ability to accumulate the most important patents concerning devices able to create, detect, or manipulate the airwaves, Isaacs blew his chance to acquire Armstrong's patent at what would have been a great bargain. Soon after Marconi's rejection, AT&T paid Armstrong $50,000 for a nonexclusive license to his invention, planning to use it to extend the distance long-distance phone calls could travel in telephone wires. The scientific paper Armstrong published describing the workings of his invention forced his fellow engineers to junk several scientific "laws" that purported to define the limits of what the airwaves could do. Marconi and Isaacs finally agreed to purchase a nonexclusive license to Armstrong's device two years after reading Sarnoff's memo, paying much more than they would have had they listened to their chief inspector from the start.

David Sarnoff also helped his new friend get a bit of revenge. In his role as the secretary of the Institute of Radio Engineers, Sarnoff supported a movement among his fellow radio experts to award the institute's first-ever Medal of Honor in acknowledgment of the epoch-making importance of Armstrong's discovery. Captain Armstrong had shipped out before he got a chance to pick up the medal itself, but news of the unprecedented honor sealed the twenty-six-year-old's reputation as the greatest radio inventor of the age.

By the time Armstrong arrived at the Marconi headquarters in London, most of the engineering staff had left for the day. Fortunately, British army captain H. J. Round, the former personal assistant to Marconi and a talented engineer whose work Armstrong knew and admired, was working late. (Among other discoveries, Round had been the first to identify the "light-emitting diodes" that would prove of great use to future generations.) The two balding army captains soon fell into an excited discussion of the latest wireless advances and their usefulness to the war effort. Captain Round, who was leading several secret projects for British signals intelligence, had to skip some confidential details of the work going on at Marconi House, but over dinner he spilled enough information for Captain Armstrong to pick up on the British strategy for turning the invisible waves into tools of war.

After three years of fighting, Captain Round explained, the Germans had become overly confident in their ability to use short-range wireless communications without the messages being intercepted. He didn't need to explain the principal reason for that complacency. Armstrong's amplifier only worked with the sort of giant airwaves the Marconi stations sent across the Atlantic, waves that stretched well over a kilometer from crest to crest. Any shorter waves that did trickle in from across the trenches were impossible for eavesdropping enemies to amplify—or so the Germans thought. Little did they know that Captain Round already had succeeded in modifying Armstrong's amplifier so that it could boost the power of waves as short as two hundred meters. Nor did they suspect that the British Admiralty had been using Captain Round's modification to track the location of warships.

At Marconi stations along the English Channel, antennas meant for catching wireless telegraphs had been turned into tools for hunting the German fleet. By boosting the two-hundred-meter waves by means of Round's version of Armstrong's amplifier, the British could analyze tiny shifts in the signals and triangulate the location of the ship sending the message. Already that information had led to World War I's largest naval battle. On May 30, 1916, well before the United States had entered the war, Round's sensors had detected something unusual: the German flagship had moved a few miles north, leaving the safety of its fortified naval base and anchoring in the Jade River, where it could quickly put to sea. British admiral John Jellicoe, who had been waiting for his chance to meet the smaller German fleet head-on, immediately ordered his twenty-eight battleships and assorted smaller warships out to sea.

Despite Captain Round's well-timed warning and Admiral Jellicoe's larger fleet, the resulting sea battle ended in a draw, with neither side losing a battleship. Still, the Battle of Jutland, as it came to be known, would keep the German fleet bottled up for the rest of the war and establish British signals intelligence as a vital source of battlefield information.

The innovative work of Captain Round had another effect as well. By the time Captain Armstrong sailed for France, he had started thinking about a new capability of invisible electromagnetic waves, which could be used not just to communicate with distant people but also to locate and track distant objects.

ARRIVING IN PARIS, Armstrong was hit by an immediate sense of déjà vu. Once again, he was greeted by radio engineers thrilled to meet the man who had invented the feedback circuit that made vacuum tube amplifiers and transmitters possible. And once again, he soon met an engineer eager to show off his system for modifying Armstrong's amplifier to boost the power of shorter waves. The French engineer Marius Latour used a slightly different method than Captain Round had employed to achieve roughly the same results, amplifying waves as short as two hundred meters. Latour's circuitry also shared the drawbacks of Captain Round's. Built with scarce components and fiendishly hard to operate, the device could never serve as an affordable and reliable tool for intercepting German airwaves.

Watching German planes and zepplins above Paris, Armstrong returned to the puzzle of how to track physical objects at a distance. Bouncing a wave off an object and trying to detect the reflection was theoretically possible, but radar, as it would come to be known, remained far beyond the limits of wireless technology in 1917. Planes, unlike battleships, rarely carried their own radios, so repeating Captain Round's Battle of Jutland trick couldn't work. The engines of enemy planes did contain traditional sparkplugs, however, and just like Heinrich Hertz's original spark-gap transmitter, those sparks released a burst of electromagnetic waves each time they fired. Might it be possible, Armstrong wondered, to turn the German sparkplugs into unwitting homing beacons?

The main hurdle was obvious—and familiar. A sparkplug's waves aren't nearly long enough to amplify, even with Round or Latour's machines. While Armstrong considered both those amplifiers brilliant pieces of work, their shared limitations reinforced his belief that incremental improvements to his last great discovery could go only so far. Or, as he put it: "The amplification of waves shorter than any ever even contemplated, quite insoluble by any conventional means of reception, demanded a radical solution."

Instead of grabbing a short wave and trying to amplify it directly, Armstrong began to build a machine that would first slow down a rapid oscillation, lengthening the distance from crest to crest. Once he had stretched the short wave into a longer wave, he could feed it into his original amplifier to boost its power, then reverse the first step to shrink the long waves back to their original length. Voilà—an amplified short wave.

Armstrong's original goal—tracking German aircraft in flight—never panned out. However, his wave stretchers let Allied forces eavesdrop on a variety of secret communications, including orders being sent to troops in the trenches and between German warships. By 1918, Captain Armstrong could be found in the skies above Paris, watching the German's long-range artillery pound the city below while he tweaked his new amplifier to intercept signals from the unsuspecting Germans in the distance. After the war, a grateful France pinned the Legion of Honor on Captain Armstrong, the U.S. Army promoted him to major (a rank he fancied so much that he decided to keep it as part of his name upon returning to civilian life), and when he returned to the United States the Institute of Radio Engineers held a ceremony to award him its first Medal of Honor for his amplifier.

FOR ARMSTRONG, returning to New York in 1919 also meant an opportunity to reconnect with David Sarnoff and catch up on a civilian wireless industry that had been remade by the war.

Armstrong arranged a meeting at his temporary laboratory at the City College of New York, eager to show his friend his wartime discovery and curious for Sarnoff to share his wartime secrets as well. Throughout the war, Sarnoff had made regular visits to Washington to educate the military brass on the technology's capabilities. His talent for explaining technology in nontechnical terms had won him many admirers among the admirals, who also appreciated his realistic estimates of how fast the Marconi Company could scale up production of different types of wireless gear. Those admirals had been shocked when they experienced, firsthand, the yawning gap between Guglielmo Marconi's promises and what his wireless stations actually could do. The realization had spurred a military investigation into the real abilities of the transatlantic telegraph system, which concluded that the only way to build a reliable wireless telegraph system was to combine dozens of different components patented by many different inventors.

Thanks to Godfrey Isaacs's tireless efforts over the previous decade, the Marconi Company had amassed the largest collection of these patents. Still, many had eluded his grasp. Rival companies, including General Electric and AT&T, and individual inventors (most notably Major Armstrong) controlled critical patents. The result, the navy concluded, was that no one company could build a state-of-the-art wireless communications system using all the best components without being sued. Meant to promote progress, the patent system had paralyzed it instead.

Using its wartime powers, the U.S. Navy solved this problem in a single stroke. American manufacturers of wireless gear could use whatever components and manufacturing tricks that worked best, the navy proclaimed, regardless of intellectual property rights. The companies would not be sued, since the military would indemnify its suppliers from claims of patent infringement.

Though Armstrong had missed out on making money from his own patents during the war, even he applauded the navy's severing of the patent system's Gordian knot. When a government investigator sought him out to ask if the drastic step had been necessary, the inventor quickly agreed: "It is absolutely impossible to manufacture any kind of workable apparatus without using practically all the inventions."

The impact of the navy's patent busting was dramatic. According to an extensive study by the Federal Trade Commission, "The Marconi Co. of America, from the time of its inception until its high-power stations were taken over by the government, had never been able to give a reliable trans-Atlantic radio service." Suspending patent rights changed that, allowing American Marconi and the U.S. Navy to succeed in building "the first high-power station on the Atlantic coast which transmitted messages continuously and reliably."

While the navy's takeover had proven to be an unexpected blessing during the war, once peace was declared, the military moved to maintain its control of the wireless industry indefinitely. Having grown used to controlling the airwaves during the war, many admirals didn't care to give them back. They particularly objected to the idea of returning control not just to a private company, but to a private company that was controlled by a British firm. As a battle in Washington brewed,

David Sarnoff became a constant presence on the train from New York to Washington. Like Armstrong and most engineers, he knew that simply returning to the patent paralysis of the prewar years made no sense. At the same time, he testified to Congress, resorting to permanent military control would rob the industry of the flexibility and entrepreneurship needed to get the next wireless advance to market.

Instead, Sarnoff threw his support behind a plan to form a single American company that would be free of British Marconi control. To solve the patent problem, the new company would collect every wireless patent into a single pool, which it would agree to license to all radio manufacturers at a reasonable rate. Faced with no viable option for keeping control of American Marconi, Guglielmo Marconi and Godfrey Isaacs cut a deal to sell British Marconi's stake in the subsidiary for $3.5 million, roughly equal to its market value at the time. Sarnoff took a lead role in negotiating the ownership stakes that each contributor to the patent pool would receive in the new company. The group included AT&T, Westinghouse, General Electric, and the United Fruit Company, which owned a few patents thanks to the wireless telegraph networks it had built on its Central American banana plantations. After that, the only real job was picking a name for the new company. On February 29, 1920, the Marconi Wireless Telegraph Company of America ceased operations. In its place, the Radio Corporation of America was born.

For the first time, a wireless telegraph company seemed positioned to finally achieve Guglielmo Marconi's eighteen-year-old dream: building a wireless network capable of challenging the global web of telegraph cables. And yet—just as his company faced that thrilling prospect—David Sarnoff's attention shifted to something else entirely.

The value of the airwaves had grown far beyond anyone's expectations, and now RCA's new commercial manager believed he had found a way to multiply it once again.

CHAPTER 7

SARNOFF

New Jersey—1921

JACK DEMPSEY LANDED A LEFT HAND TO HIS OPPONENT'S NOSE, breaking it. From every direction, David Sarnoff heard a cacophony of cheers and gasps as the European heavyweight champ staggered backward.

The bloodied French fighter, Georges Carpentier, looked almost puny next to the American heavyweight. The ring announcer had tried to fudge the difference, padding Carpentier's official weight to get him in the ring at 175 pounds while shaving a few pounds off of Dempsey to get him down to 188. No amount of creative mathematics could disguise the size mismatch now, as the larger American circled his smaller foe, looking for another opening. He found it, slipping a straight right hand past Carpentier's guard that sent the Frenchman tumbling into the ropes and nearly out of the ring.

"The greatest crowd that has ever invaded New York City for a sporting event," as one newspaper described it, had been pouring into Manhattan for the past week. Most of the super-wealthy arrived in private train cars. Other came by yacht, including the actress Mary Pickford, who sailed in from Hollywood by way of the Panama Canal. The city's best hotels were booked solid with visiting fight fans: three hundred rooms at the Plaza, five hundred at the Waldorf Astoria, eight hundred at the Biltmore.

Near the ring Sarnoff could spot movie star Douglas Fairbanks, singer Al Jolson, industrialist Henry Ford, and a dozen U.S. senators. Filling the seats in between were various Rockefellers, Guggenheims, Vanderbilts, Astors, and Roosevelts, all dressed in their summer finest: ties, light jackets, and white straw hats to shield their faces in case the midafternoon sun emerged from the clouds now hanging over the arena. Since New York had strict laws governing boxing matches, the two fighters and their ninety-one thousand fans had all come here, to a previously anonymous strip of New Jersey known as Boyle's Thirty Acre Woods, where a giant wooden arena had been built to host the fight.

Tex Rickard, the fight's promoter, had billed his championship bout as "The Battle of the Century," a label that in 1921 had not yet grown stale from overuse; and that many people considered an understatement in any case. Two hundred reporters from newspapers around the globe were at ringside, another five hundred from lesser-known publications in the cheap seats. As the roar of the crowd faded, Sarnoff could hear the clackety-clack of typewriters mixed with the chatter of telegraph keys. Four dozen cables operated by the Postal Telegraph Company ran into the stadium, along with another twenty-five from rival Western Union. To speed delivery of the news, some of the transatlantic cables had been temporarily rewired to offer a direct feed from ringside in New Jersey to Paris and London. As the fighters sat in their corners, waiting for the bell to start the second round, news reports of Dempsey's dominating start were already being read by anxious fight fans around the world.

Supporters of the Frenchman had insisted that despite appearances, the brute strength of the larger Dempsey would crumble when faced with a smarter, more technically skilled fighter. A twenty-two-year-old newspaper writer for the *Toronto Sun,* Ernest Hemingway, made the case this way:

> If we are to believe the experts and some of the editorial writers in U.S. newspapers, it is practically suicide for Georges to climb into the same ring with Jack. Dempsey will hit him once and it will all be over. Dempsey is the greatest heavyweight of all time. It looks bad for Carpentier.
>
> Experts are all victims of "Championitis." Whoever happens to be the titleholder is the greatest fighter of all time. Thus they write reams about the wonderful superman that is Dempsey.
>
> It is bunk and twaddle of the worst kind.
>
> Jack Dempsey has an imposing list of knockouts over bums and tramps, who were nothing but big slow-moving, slow-thinking setups for him. He has never fought a real fighter.

The bell for second round rang and Carpentier approached his larger foe cautiously, hoping to find a way to prove Hemmingway right.

A TITLE FIGHT was one place where David Sarnoff felt a bit out of place, even if he didn't show it. In his twenty years in America, he had displayed virtually no interest in popular entertainment, invariably choosing an evening at the office to a night at the fights. Today, however, was a day he had been looking forward to just as eagerly as the rabid fight fans surrounding him—perhaps even more.

The RCA board of directors had just promoted the now twenty-nine-year-old from commercial manager to general manager, a title that put him in charge of all the young company's operations and rewarded him for his dogged efforts to establish international wireless telegraphs as a legitimate competitor to the submarine cable cartel. RCA's new transatlantic wireless network, which Sarnoff oversaw, had just launched its attack on the transatlantic cables. RCA's first link connected New York to London, showing that the waves could compete with wires on the world's busiest cable route; its second connected New York to Norway, demonstrating how the waves in the air would create direct links to places cables could only reach after many expensive hops through multiple countries' telegraph networks. The cable rates to London, pegged at twenty-five cents a word for the last thirty-five years, suddenly faced a competitor moving "radiograms" for just seventeen cents a word. (As of a month earlier, "Marconigrams" had been rebranded as radiograms.) The cables charged thirty-five cents a word to reach Scandinavia. RCA charged twenty-four cents.

Sarnoff knew that stock traders, whose business was the richest part of the cable business, needed to send and receive their messages quickly, so he focused on making his network cheaper, faster, and more reliable. RCA's airwaves could carry a London stock trader's order over the Atlantic at the speed of light, banking off the atmosphere before striking a seven-mile-long antenna that RCA had strung up on Long Island. From there the message got a power boost from Armstrong's amplifier and used the energy to zip down a sixty-five-mile-long telegraph cable running into New York City.

To speed things along, Sarnoff even did away with his old job, replacing human telegraph operators with the new generation of machines that could turn words into airwaves and airwaves back into words quicker than the fastest fist. On the sending side, new gadgets used a prepunched strip of paper to trigger a blazing stream of dots and dashes at speeds topping fifty words a minute. On the receiving side, special typewriters sensed the incoming waves and "demodulated" the interruptions into dots and dashes printed on a strip of paper. Besides being faster, the machines printed records of messages that helped eliminate human error and reduce customer complaints. Rather than forestalling or bemoaning the inevitable end of his old profession, David Sarnoff worked to hasten and celebrate its demise. "Radiograms travel at the speed of light and from the moment of transmission in Europe to direct typewriter reception in New York City, no hand relaying is involved," RCA's marketing department boasted. The era of the telegraph operator was over. The age of mechanical modulators and demodulators—or "modems," as the increasingly ubiquitous machines would come to be known—had begun.

It had taken two decades, several major scientific breakthroughs, and tens of millions of dollars, but the wireless telegraph was finally living up to Guglielmo Marconi's grand promises. Using new transmitters, Sarnoff could skip messages right over the United Kingdom, which had controlled the instantaneous movement of information around the world for the previous six decades. RCA's new direct link to Germany had already proven so popular that the company was opening a second station to double capacity. A Hawaii-to-Japan link was also up and running, and soon RCA's waves would connect California to Hawaii. Construction was also under way for new wireless stations linking RCA's central radio office in New York with Eastern Europe and South America, tapping still more new markets.

For David Sarnoff and his boss in New York, Owen Young, the ultimate proof was in the numbers. By the close of 1921, RCA's first full year in operation, its airwaves would collect $2.1 million for moving eighteen million words, or nearly 20 percent of the transatlantic market for instant communication. With the construction boom in new stations, that growth promised to continue for years to come.

And yet, even as his moment of triumph approached, the twenty-nine-year-old Sarnoff had moved on to the new, far more ambitious dream that had brought him to the fight today.

As with most major news events in the early twentieth century, there was no easy way to instantly share information with all the people who were interested. At this particular moment, crowds were gathered near telegraph stations around the world, waiting for news to arrive and be announced. In New

York's Times Square, a crowd of twenty thousand craned their necks to read the updates posted on the *New York Times*' marquee. In Paris, people particularly anxious for word of their national champion had devised multiple ways of sharing the telegraphed updates. One newspaper had set up loudspeakers to broadcast updates to the crowds outside its office; in another suburb telegraph operators were posting each round's update on billboards. The French army had set up a special system for the big announcement. Once word of a winner came by the telegraph, a small squadron of planes would take to the air. They would then broadcast a single bit of information to the city below. A white flare would signal a Dempsey victory. A red flare above Paris would ignite a citywide celebration.

The archaic methods of distributing information had become Sarnoff's new obsession. Telephone and telegraph wires did a great job of delivering messages to individual people, but neither the telegraph nor the telephone industry was any good at rapidly disseminating information to large groups of people. Sarnoff planned to change that. For years, he had been toying with a new use for the airwaves—a second wireless app to go along with the wireless telegraph. After he saw Edwin Armstrong's airwave-lengthening invention a year earlier, those fuzzy ideas began to solidify. Now, as he stood among ninety-one thousand fans watching Jack Dempsey pummel his opponent, Sarnoff knew he was onto something huge.

Sarnoff had conducted his first experiment in broadcasting in May of 1914, four months after seeing Armstrong's amplifier in action at Belmar. On a ship cruising around New York, he impressed his fellow passengers by playing music from a record player stationed at the Marconi station atop Wanamaker's department store. It had been a neat trick—albeit one that seemed pointlessly complicated compared to simply bringing the phonograph on the boat.

A year later, Sarnoff had worked out enough of the problems to turn his "radio music box" idea into a formal pitch to his bosses at the Marconi Company: "I have in mind a plan of development which would make radio a 'household utility' in the same sense as the piano or phonograph. The idea is to bring music into the house by wireless. While this has been tried in the past by wires, it has been a failure because wires do not lend themselves to this scheme. With radio, however, it would seem to be entirely feasible."

Sarnoff continued:

> The receiver can be designed in the form of a simple "Radio Music Box" and arranged for several different wavelengths, which should be changeable with the throwing of a single switch or pressing of a single button.
>
> The "Radio Music Box" can be supplied with amplifying tubes and a loud speaking telephone, all of which can be neatly mounted in one box. The box can be placed on a table in the parlor or living room, the switch set accordingly and the transmitted music received. There should be no difficulty in receiving music perfectly when transmitted within a radius of 25 to 50 miles. Within such a radius there reside hundreds of thousands of families; and as all can simultaneously receive from a single transmitter, there would be no question of obtaining sufficiently loud signals to make the performance enjoyable.

His Marconi Company bosses had ignored that memo, but Sarnoff continued to mention the idea regularly in the years that followed. Today, the newly promoted general manager of RCA was going to get to try out his idea on a grand scale.

DAVID SARNOFF FIRST LEARNED the value of a good publicity stunt from Guglielmo Marconi, and thanks to the fierce interest in today's fighters, the title bout offered the publicity opportunity of the century.

Beyond the obvious stakes in a fight between a European and American heavyweight champion, public interest in the fight centered on the wildly different personalities and reputations of the two men in the ring. As young men, both fighters had been prodigies with their fists, but that was where the similarities ended. Growing up in Colorado, Jack Dempsey fought his first bouts in saloons, where he won bar bets by taking on much older and larger opponents. When America mobilized for war in 1917, Dempsey avoided the draft by claiming that he was his family's sole means of support—a claim his embittered ex-wife publicly denied after the war ended. The jury in his federal draft evasion case found Dempsey not guilty, but the rest of the nation was less forgiving. Wrote sportswriter Grantland Rice: "It would be an insult to every doughboy who took his heavy pack through the mules' train to the front-line trenches to go over the top to refer to Dempsey as a fighting man."

The suave French fighter, on the other hand, seemed tailor-made for the role of the scrappy hero. A pro by age fourteen and European welterweight champion at age seventeen, Carpentier climbed rapidly through the weight classes, winning as he went. The Great War temporarily halted his rise, but only bolstered his heroic reputation. (He ended his stint as a pilot in the French army having earned two of the country's top military honors.) Many American boxing fans mocked Carpentier. Nicknamed, "The Orchid Man," after the flowers he often wore on his lapel, Carpentier struck many as too effete to take on the Manassa Mauler. His training regimen, which included shadowboxing with trees in Parisian parks, aiming for specific leaves in order to sharpen the accuracy of his punches, didn't help. Jack Dempsey preferred to train by knocking out the teeth of his sparring partners.

To the relief of Hemingway and the tens of thousands of Carpentier backers in the crowd, midway through the second round, Carpentier managed to land his first solid blow. After the referee stepped in to separate the two fighters from a clinch, Carpentier surprised the larger fighter. Instead of backing away, he stepped in, sneaking a left hook over Dempsey's guard. Though not enough to knock Dempsey down, the blow briefly slowed the larger fighter. A few seconds later, the lithe Frenchman feinted with his left and followed with his famous right hand, catching Dempsey square in the face. Here, it seemed, was the blow that the Frenchman's fans had prayed for, and they jumped to their feet with a deafening cheer. Suddenly David Sarnoff had something to worry about besides his complicated network, as the newly constructed wooden arena swayed with the force of the exultant fans.

At ringside, one member of the press seemed out of place: a brown-haired man in eyeglasses bellowing a description of the action into a telephone. Andrew White's regular job was editing the magazine *Wireless World*, but today, if things were working as he and Sarnoff had planned, his blow-by-blow description was traveling over a phone line leased from AT&T for two miles to a train station that he had converted into a temporary broadcasting station. There, Pierre Boucheron, an RCA man chosen for his charm and vocal timbre, repeated White's descriptions into a microphone. Finally, a 3,500-watt transmitter fed a signal into a long wire that Sarnoff's men had strung between a steel tower and the train station's clock tower. With luck, that wire was now emitting 1.6-kilometer-long waves that were bouncing of the ionosphere and beaming down on every city and town within two hundred miles.

Technically, nothing about Sarnoff's broadcasting experiment was revolutionary. Devices that could turn sound waves into electromagnetic waves, known as microphones, had been around for half a century. A Welsh-American named David Hughes had invented the first microphone, and Thomas Edison independently built a very similar device, which Alexander Graham Bell and other telephone pioneers then used as a central piece of their system for sending the human voice over wires. In 1906, a Canadian named Reginald Fessenden sent voice-carrying electromagnetic waves through the air for the first time. Ten years later, AT&T teamed with the U.S. Navy to demonstrate "radio telephone" calls across the Atlantic. Still, the idea of using a radio telephone to do anything but make phone calls shocked many people. The head of the American Radio Relay League, the country's leading organization of radio operators, predicted the boxing broadcast would travel no more than one hundred miles and even then would be audible only to a listener wearing headphones.

As far as AT&T was concerned, the best the airwaves

could do was offer an alternative to running telephone wires to remote or hard-to-reach locations. The phone monopoly had invested nearly $1 billion (in 1920 dollars) into its ever-expanding web of landlines, a scale that allowed the company to increase its network's reach, reliability, and sound quality while gradually dropping the price of a call. In a head-to-head fight, AT&T's Theodore Vail pointed out, telephone wires possessed immutable advantages over the waves traveling through the air. When a wireless message blasted out of a radio transmitter, its waves headed every which way. The waves that happened to reach a distant telephone receiver represented a minute fraction of the total energy emitted by the transmitter, the rest going to waste as it spread out over vast areas, unnoticed and unvalued. Vail declared that the airwaves were therefore inherently inefficient compared with his network of wires on the ground, and lacked privacy to boot: "A wireless telephone talk is a talk upon the housetops with the whole world for an audience."

The whole world for an audience—that was exactly what David Sarnoff found so intriguing.

While middle- and lower-income Americans had been priced out of the best seats at the Dempsey-Carpentier championship bout, their passion for the long-awaited event was perhaps even more rabid, as Sarnoff was reminded every time the cheaper sections of the stadium erupted in cheers. Tex Rickard, the promoter, had built the wooden stadium in Jersey City, New Jersey, just for this fight, designing a low-rise octagon with plenty of room for all income brackets. Seat prices started at $50 for a prime spot at ringside, then fell as the distance from the ring increased, to as little as $5.50. The low prices and wide distribution had allowed many poorer fans to scrape together enough for a train ticket to New York and a seat at Rickard's stadium.

Sarnoff liked Rickard's pricing system so much that he wanted to extend it to the entire country. The ninety-one thousand fans in the stadium were just a tiny fraction of the people who would pay to get live reports of the action. If he could distribute a blow-by-blow account to the tens of millions of people who wanted to hear the fight, he could charge them each a nickel and still make a fortune.

The technology to build Sarnoff's radio music boxes was still in its infancy and consumer demand for such a fanciful product nonexistent. With no radio listeners to hear them, no radio stations were on the air. Sarnoff was counting on Georges Carpentier and Jack Dempsey to solve this chicken-or-the-egg problem. RCA had teamed up with the National Amateur Wireless Association, with RCA agreeing to get the fight on the air and the Amateur group signing up to solve the receiver side of the problem. A month earlier, that group sent out its pitch for volunteers: "The greatest international sporting event in the history of the world, the Dempsey-Carpentier boxing match on July 2nd, will be voice-broadcasted from the ringside by radiophone on the largest scale ever attempted," the letter began. Amateurs were welcome to rent out auditoriums and host large groups, with any ticket sales to be donated to charity.

While David Sarnoff hoped to debunk Theodore Vail's assumption that one-way radio telephone messages had no value, his broadcasting plan traced its lineage to one of Vail's fundamental insights into the information-moving business. According to Vail, when a network cuts the price of communicating, it often spurs so much new business that the network's profits don't fall, they rise. When the first transatlantic cable opened for business in 1866 it charged $10 per word to move a word across the ocean. The cable monopoly soon realized that it could make more money charging $5 a word, since the lower price more than doubled traffic on the cable. More price cuts followed, to $2.50 then $1.25, with profits rising each time.

Human beings have a boundless desire to connect with one another. The more they can afford, the more they will connect. Profit, Vail preached, "can be produced in two ways; by a large percentage of profit on a small business or a small percentage of profit on a large business"—and the second option was almost always far superior for everyone involved.

Whether through osmosis or conscious imitation, David Sarnoff had absorbed Vail's philosophy and was now using it to attack a market the phone company considered its own: delivering the human voice. If a transmission could be picked up by thousands or even millions of listeners, the cost could be spread out among them, thus slashing the price of each individual connection. Unlike a telephone connection, the communication could flow in only one direction. But when it came to events like today's fight, a one-way broadcast of information was all anyone wanted anyway. Inventing a new wireless technology like Armstrong's amplifier was only one way to radically increase the value of the airwaves, Sarnoff suspected. Inventing a new wireless app might work too.

JACK DEMPSEY, WHO among his other talents knew how to take a punch, regrouped after absorbing Carpentier's right hook, staying out of his range in the closing seconds of the round. As the third round opened, the scowl was back on Dempsey's face, the momentum back in his corner. Though the crowd didn't recognize it yet, it was the French fighter who had been badly hurt in the second round. Not only had the Manassa Mauler weathered Carpentier's big, right-hand punch, the blow broke Carpentier's thumb, crippling his best weapon. Now the smaller challenger was back to dancing just out of Dempsey's reach, a cat-and-mouse game that continued to frustrate the champ.

With less than a minute into the fourth round, Dempsey landed a shot that sent Carpentier to the canvas. As the referee began his ten count, Carpentier lay quietly until the count of nine, then sprang to his feet and charged at Dempsey. Whether or not he was impressed with the French fighter's bravado, Dempsey knew how to repay it, connecting a solid right to his incoming chin. This time there was no playacting. Carpentier went down, and at one minute and sixteen seconds into the fourth round, the count reached ten and the stadium erupted in cheers, wails, and the clattering telegraph keys.

In theaters and auditoriums up and down the East Coast, crowds of fans were thrilled with the experiment. Sarnoff later claimed that three hundred thousand listeners had heard the broadcast. *Wireless Age,* a publication started by the Marconi Company and now run by RCA, reflected the enthusiasm Sarnoff's bosses had for his brainchild:

> Radio has had its triumphs. Great distances have been spanned in the past, nations and continents have been connected; even the voice has been carried across the sea. But everything in the past record of wonders but adds to

the luster of this latest amazing demonstration of broadcasting a voice to the largest audience in history.

There is now an insistent demand that the idea be kept alive, that large scale broadcasting to audiences, through the amateurs, be expanded to include voice descriptions of baseball games, all sorts of sporting events, speeches by noted men, lectures and every imaginable form of musical entertainment. Enthusiasm has reached high pitch, as the doubters have been silenced.

CHAPTER 8

HOOVER

Washington, D.C.—1924

HERBERT HOOVER WAS NOT QUITE SURE WHAT TO MAKE OF THE young executive at the lectern. David Sarnoff was bold, certainly. The thirty-three-year-old, currently addressing an audience composed largely of local radio-station owners, had spent the last several minutes cataloging the many problems posed by the country's reliance on those station's short-range radio service. Having finished offending the local station owners, he was now busy scaring them.

Using RCA's latest vacuum tube transmitters, Sarnoff declared, new "super-power radio stations" would begin blasting towering waves at the ionosphere. These mighty waves would bounce back to earth with enough energy to blanket a dozen or more states with an AM radio signal that consumer radios could pick up with ease. Locating these new stations in the country's major cities, Sarnoff added, would give them access to top-flight musicians, opera singers, comedians, and actors, another feature the local AM stations couldn't match. Most importantly, establishing these superpower stations would ensure that every American would be able to tune in several radio stations on their dial. The Radio Corporation, Sarnoff announced, already was working on plans to build an experimental superpower station in the Northeast. "Thereafter, in close technical and practical cooperation with its associates, the system would be extended to cover every nook and corner of the United States."

Herbert Hoover did not need to look at the radio men in the audience to know how they were taking Sarnoff's promise to "add vastly to the facilities which now exist." The competitive threat the Radio Corporation's proposed stations posed to the local station owners' new businesses was self-evidently terrifying.

To be sure, the problems Sarnoff described were real, as Hoover was in a position to know as well as anyone. Now in his fourth year as secretary of commerce, Hoover had taken over the often thankless job of deciding which radio stations deserved to be on the air, parceling out licenses to beam out radio-carrying electromagnetic waves of a specific length. So far, he had granted licenses to six hundred radio stations that had sprung up in the past three years, and more were clamoring for their own airwaves every week. Most of these stations were low-budget affairs, relying on relatively low power signals that could spread a radio signal across a city or county, but not much farther. That left large stretches of the country, including vast swaths of rural America, with no available radio programs at all—a problem that David Sarnoff now proposed to solve.

BEFORE HE BECAME A POLITICIAN, Secretary Hoover had spent his career in the mining industry, first as an engineer and later as a mine owner himself. Rich enough to retire at age forty-five, Hoover turned to politics. With the 1920 presidential election coming up, he decided that running for president would be a good way to start.

Unsurprisingly, his rivals for the Republican nomination mocked Hoover's lack of political experience and presumptuousness. The would-be president's last elective office had been a stint as treasurer of Stanford University's student body. His most significant nonelected role in public life consisted of running a little-known federal agency tasked with helping reduce domestic food consumption during World War I. A humiliated Hoover stumbled to a fourth-place finish in a race won by Warren Harding, a heretofore undistinguished senator with a modest intellect and dubious morals.

President-elect Harding salvaged Hoover's nascent political career by offering him a spot in his new administration, albeit as head of the cabinet agency widely considered Washington's dullest and least important. As far as Hoover was concerned, however, the commerce secretary job was just what he needed: a chance to prove to voters that he could engineer and administer wise government policies. The radio boom had offered the science-minded commerce secretary an obvious place to begin. The next time he ran for president, Herbert

Hoover intended to boast more impressive political accomplishments than auditing a university athletic department's books and developing a wartime ad campaign for "Meatless Mondays."

Even before Hoover took his oath of office alongside the rest of President Harding's cabinet in 1921, the neophyte politician had reason to worry about the incoming administration's ethics. Harding's first pick for interior secretary, an Oklahoma oilman named Jake Hamon, had a particularly shady reputation—and that was before his longtime mistress, irate over his plan to move to Washington without her, shot him in the stomach. The ensuing murder trial wrapped the Harding administration in scandal even before the new president took office. The real trouble, however, began when President-elect Harding replaced the dead man with an equally dubious character, New Mexico senator Albert Fall.

As interior secretary, Fall was responsible for leasing government land to oil companies, and he quickly leased several oil fields at prices far below what the government could have received at auction, pocketing large bribes in exchange. The graft might have gone unnoticed had his increasingly opulent lifestyle not aroused the jealousy and suspicion of his former Senate colleagues. Soon the whole country knew the details of Fall's thievery as well as the name of the Wyoming oil field that he had traded for a $400,000 bribe: Teapot Dome. Convicted of bribery and conspiracy, Albert Fall became the first cabinet secretary in United States history to be thrown into jail for crimes in office. In the wake of that famous scandal, Secretary Hoover set about designing a system that would protect the American airwaves from the same sort of corruption.

IN MANY WAYS, David Sarnoff was an easy man for the commerce secretary to like. The former engineer and business executive appreciated Sarnoff's unusual combination of technical and financial acumen. The thirty-three-year-old also had a knack for translating complicated science into plain English, an increasingly valuable skill in Washington as Congress considered new laws governing an industry about which they were thoroughly ignorant. The most compelling endorsement of David Sarnoff came from Hoover's friend Owen Young, the chairman of both General Electric and RCA, its partially owned subsidiary. Sarnoff's boss not only raved about his employee's business savvy, he also clearly felt a personal bond with the younger man. Still, Hoover couldn't shake his lingering doubts.

In addition to his conflict with the local radio-station owners, Sarnoff had also started a bitter war with the country's most powerful communications company. The men who ran American Telephone & Telegraph following Theodore Vail's death in 1920 continued to subscribe to Vail's theory that since radio involved moving the "speaking vibrations" (as Vail called electromagnetic waves used to carry the human voice), it was a natural extension of the phone system. In 1924, the only way to move a radio program between a radio station in Boston and a station in New York was over a phone line, since rebroadcasting a radio program picked up over the air introduced unacceptable distortion. Therefore, according to the phone company's view of the world, any radio network should be designed around their wires.

In an internal memo, an AT&T vice president put it this way: "We have been very careful up to the present time not to state to the public in any way, through the press or in any of our talks, that the Bell System desires to monopolize broadcasting; but the fact remains that it is a telephone job." Hoover had not seen that candid assessment, disclosed as part of a later lawsuit, but the phone company's attitude was apparent just the same. Nor did he have any doubt about the phone giant's negative reaction to Sarnoff's new proposal, since superpower stations would cut the phone company out of the radio business altogether.

The roots of the feud between the Radio Corporation of America and American Telephone & Telegraph traced back to Owen Young's decision to assign Sarnoff the job of negotiating with AT&T during the formation of the Radio Corporation in 1919. The main part of that deal was unremarkable: in exchange for contributing several patents to RCA's patent pool, the phone company received a 10 percent ownership stake in the new company. The controversy sprang from a side agreement that Sarnoff negotiated with AT&T as part of the contract, an ancillary provision that few people considered important at the time. Under its terms, AT&T and RCA divvied up the different pieces of the radio manufacturing business. AT&T retained the plants and patents needed to build radio transmitters; RCA kept the radio receiver side of the business. This seemed like a square deal in 1920, a time when every radio telegraph and radio telephone came with one transmitter and one receiver. By 1924, however, it was clear to Ma Bell (as the phone monopoly was nicknamed) that David Sarnoff had played her for a chump.

This year, RCA's sales of consumer AM radios were on course to top $50 million—with no end to the astounding growth in sight. Meanwhile, AT&T's side of the bargain left it with a market of just six hundred radio stations—most of which chose to ignore the company's patents and construct their own transmitters. (Phone-company lawyers sued, igniting a public firestorm over AT&T's efforts to "monopolize the air" and forcing the company to give up on collecting even the small amount of revenue to which it was legally entitled.) Despite this series of embarrassments, the phone monopoly showed no signs of giving up its quest to control the new industry.

To retaliate against Sarnoff and RCA, AT&T connected its flagship radio station, New York's WEAF, to a series of stations in smaller towns. Just this month the phone company demonstrated the reach of its new radio network by letting President Coolidge give an address over twenty-two AM radio stations at the same time, an unprecedented feat. When Sarnoff tried to negotiate a deal to allow RCA to begin forming its own network of radio stations connected by long distance telephone lines, AT&T refused to take his business.

At the same time, AT&T's collection of radio stations began trying out a new business model that had been rejected by many in the industry, including David Sarnoff and Herbert Hoover. To fund the stations' operating expenses, the phone company sold access to its radio network by the minute, just like it sold telephone service. WEAF first tried this in 1922, pocketing fifty dollars for letting a velvety-voiced pitchman spend ten minutes extolling the virtues of a new suburban housing development on the air. ("Friend, you owe it to yourself and your family to leave the congested city and enjoy what nature intended you to enjoy. Visit our new apartment homes in Hawthorne Court, Jackson Heights, where you may enjoy community life in a friendly environment.")

AT&T called this innovation "toll broadcasting"; Herbert Hoover called it "direct advertising," and in his own speech to the conference this morning, had declared it "the quickest way to kill broadcasting." Ignoring the success of AT&T's experiment, which led to a rush of interest from American Express, Macy's, Gimbel's, Wanamaker's, and a parade of other new radio advertisers, Hoover hit the idea hard: "The reader of the newspaper has an option whether he will read an ad or not, but if a speech by the President is to be used as the meat in a sandwich of two patent medicine advertisements, there will be no radio left."

So far, David Sarnoff and RCA had sided with Hoover. Since RCA made its money selling radios (and licensing its patents to other radio manufacturers), it had a vested interested in boosting sales. David Sarnoff had even volunteered to write a statement supporting the ban of direct advertising and backing Hoover's plan to place a surcharge on radio sales that could be pooled to pay for the radio stations.

Hoover knew that while the local radio-station owners and phone-company representatives disliked RCA and its aggressive general manager, Sarnoff did have allies among the engineers attending the radio conference. This group certainly included Edwin Armstrong, whose bald head could be seen sticking up out of the audience. Like most experts, Armstrong was happy to debunk local radio-station owners' fears of fifty-thousand-watt superpower signals "drowning out" their local fifty-watt transmitters. The engineer's argument was simple: as electromagnetic waves travel through the air, their power drops at an exponential rate. By the time a fifty-thousand-watt signal bounces off the ionosphere and returns to earth, the radio signal it delivers will be less powerful than a local fifty-watt station whose airwaves travel directly from a broadcast antenna to a nearby listener's radio.

Straightforward as this science was, most attendees showed little interest in accepting it. Ever since the idea of increasing power to fifty thousand watts began to circulate, Hoover had received a steady series of visits from station owners opposed to the notion. They were backed by newspaper-industry representatives in the audience, who felt exactly the same way about the prospect of new competitors descending from the sky and stealing the local advertisers that the newspapers and local radio stations now divided up among themselves.

From the podium, David Sarnoff wrapped up his speech by attempting to assuage the anxiety his superpower station proposal had been causing his audience. The Radio Corporation made its money selling radios, he pointed out, and thus it benefited from having as many stations as possible on the air. "The Radio Corporation of America has every interest to encourage the maintenance of the local stations," he assured his skeptical audience. "In embarking upon super-power broadcasting, we are only following the progress already made in radio reception, and propose to increase, rather than diminish, the value of the receivers in the homes."

If Sarnoff believed this would mollify his audience, maybe he wasn't as smart as Herbert Hoover had thought. As soon as he finished, representatives of the smaller stations fanned out to voice their objections to anyone who would listen, including Secretary Hoover and the handful of reporters who had shown up to cover the conference. Many of the stations had only just begun to sell enough ads to turn a profit, and now one or more national networks with more expensive programming were going to come steal their audience away? Everywhere Hoover turned he seemed to find another angry objector. Walter Strong, the head of the powerful American Newspaper Publishers Association, protested with particular vehemence. On top of these opponents, Hoover knew that AT&T, perhaps the most powerful corporation in America, could be counted on to do everything possible to quash Sarnoff's idea.

By the time the session disbanded for the day, the widespread revulsion many of the attendees felt toward Sarnoff's idea had overshadowed the proposal itself. The next day's *New York Times* headlined its article "DELEGATES OPPOSE HIGH-POWER RADIO; EXPRESS FEARS AT WASHINGTON CONFERENCE OVER RADIO CORPORATION IDEA OF SUPER-STATIONS."

A vote to reject Sarnoff's proposal soon followed. In its place, the radio conference approved a (nonbinding) resolution declaring that: "The conference has been strongly urged to recommend the abolition of all limitation on power, but it refuses to do so. There has been no experience in this country and little anywhere else in the world with broadcasting by stations of such power. In the absence of actual knowledge of their effect or usefulness, the conference refuses to recommend any authorization of the general issue of licenses for stations of this character."

FOLLOWING THE END of his conference, Hooever turned his attention to updating the laws that govern the use of the airwaves. The newborn AM radio industry was governed by a law passed in the early days of the wireless telegraph, long before even David Sarnoff began dreaming of broadcasting as the airwaves' second commercial application. Passed in the aftermath of the *Titanic* disaster, the Radio Act of 1912 focused on bringing order to the airwaves. Without rules governing who could transmit wireless telegrams, traffic jams in the sky were inevitable. (In the days after the *Titanic*'s sinking, amateur wireless telegraph operators competed to relay details of the event, garbling everyone's messages.) Hoover's challenge was to adapt that law to help the new radio broadcasting industry, which was also struggling with its own invisible traffic jams. He viewed the 1912 law as sufficient legal authority for him to issue radio licenses to applicants of his choosing—a view not shared by many of the applicants he had rejected.

Hoover's legal troubles started when Zenith Radio Corporation, a radio manufacturer that also ran its own station, objected to the terms of its Department of Commerce license. When Hoover refused to change those terms, which restricted the station's right to broadcast to Thursday night between 10 P.M. and midnight, irate Zenith executives ignored his ruling and started broadcasting outside the allotted time slot. Hoover responded by initiating criminal proceedings against the company, accusing it of violating the 1912 Radio Act. His case quickly crumbled. The drafters of the 1912 law had blithely assumed that no new tools for harnessing shorter airwaves would be invented, and therefore gave the government the right to license only those airwaves that were at least six hundred meters long. On the shorter airwaves used by more modern equipment, a federal judge told Hoover, Zenith and anyone else could broadcast whatever they wanted.

Chaos ensued. WHAP in New York City wedged itself in between WJZ and WOR, garbling the signals of both of the popular and well-established stations. KFI and ten other western stations protested the anarchy in the air with what they called "interference hours," altering their wavelengths to intentionally overlap. For an hour West Coast radios would emit nothing but squeals, howls, garbled speech, and mangled music—after which the stations broadcast pleas to support a new radio law.

Capitalizing on the situation, Hoover teamed with Senator Clarence Dill to draw up a law to govern the airwaves in the modern age and shield it from the threat of Teapot Dome–style corruption. The chaos in the airwaves required "that Congress establish an entirely independent body to take charge of the regulation of radio communication in all its forms," Dill argued. "The exercise of this power is fraught with such great possibilities that it should not be entrusted to any one man nor to any administration department of the government. This regulatory power should be as free from political influence or arbitrary control as possible."

Senator Dill's solution of an independent commission troubled both Hoover (who wanted to keep the control over radio in his cabinet department) and Coolidge (who considered such regulatory agencies to be a hybrid of legislative and judicial functions that the Constitution had been designed to keep separate). In order to get the bill through Congress, Hoover convinced Coolidge to concede that central issue to Senator Dill. As a result, the Radio Act of 1927 created a new federal agency, the Federal Radio Commission, to oversee the airwaves. (In 1934 the commission was given jurisdiction over other types of electronic communications and a new name, the Federal Communications Commission.) Herbert Hoover considered the law his signature triumph, and its many safeguards against corruption a shining example of good government in action.

In the decades that followed, however, the well-intentioned legislation would result in something Herbert Hoover had not expected: corruption on a scale so vast that Teapot Dome would pale in comparison.

CHAPTER 9

SARNOFF

New York City—1933

WALKING ACROSS THE ROCKEFELLER CENTER PLAZA WITH Edwin Armstrong offered David Sarnoff a welcome respite from his battles with shareholders, accountants, and board members who didn't share his unshakable optimism. The sprawling office complex, located in the heart of Manhattan, had been given a holiday makeover to welcome its new tenant, the Radio Corporation of America. Officially, the tradition of erecting a Christmas tree in front of "30 Rock" had begun two years earlier, when the steelworkers building New York City's largest office complex commemorated the Great Depression's third Christmas by trimming a small Christmas tree with gum wrappers. This year, though the Great Depression continued unabated, RCA new landlords had trucked in a forty-foot spruce and illuminated it with seven hundred lights that twinkled above the two old friends' heads.

Sarnoff and Armstrong looked very different from when they had first met. Sarnoff's baby face was long gone. A,new smattering of gray hair at his temples testified to the strain of the past few years. (He wore it cut short and slicked back, in the style of the day.) While Armstrong, an avid tennis player, remained tall and thin, Sarnoff had allowed his waistline to grow, though he cloaked it well with his custom-tailored suits. Armstrong had celebrated his forty-third birthday earlier in the week; Sarnoff was set to join him in two months.

For the past twenty years, the inventor had been stopping by David Sarnoff's offices, each larger and grander than the one before. Armstrong had gotten to know the woman who was now his wife, Marion Macinnis Armstrong, during his regular visits to Sarnoff's old office on the eighteenth floor of the Woolworth Building, where she had worked as Sarnoff's secretary. None of Sarnoff's previous office upgrades, however, came close to his latest. Today David Sarnoff didn't have just a new office to show off, he had a grand complex he called Radio City, which had opened for business just a few weeks earlier.

John D. Rockefeller had cooked up the idea of building New York's largest office complex during the Roaring Twenties, just before the stock market crashed. When the Manhattan real estate market crashed soon thereafter, David Sarnoff stepped in to save him by signing the largest lease in New York history. Sarnoff's gamble seemed dangerously optimistic. RCA and NBC certainly needed room to expand, but existing Manhattan office space was cheap and landlords desperate. Sarnoff ignored this. He wasn't interested in finding a corporate headquarters fit for riding out the Great Depression—he wanted to create a home for RCA and NBC that would prepare them for the brighter decades ahead.

After persuading the RCA board to approve his proposal, Sarnoff poured his shareholders' dwindling cash into outfitting the new complex with a Marconi-esque abandon. Everywhere Armstrong looked, he could see signs of Sarnoff's grand ambitions. The National Broadcasting Company now had its own building with dozens of radio studios, each one built atop giant, felt-covered springs designed to absorb the noise that otherwise might filter in from the city streets. Radio City's state-of-the-art air-conditioning system—a technological marvel in 1933—also filtered dirty city air through spun glass fibers custom-designed by RCA engineers. Every mechanical device had been built with a backup in case a machine failed. The building generated its own electrical power on site, but if that conked out, NBC could tap into power from any of eight different backup connections to power sources around the city. Should all nine of those sources fail, perhaps in the wake of a natural disaster, the building could turn to the storage battery built into the basement that could keep the whole operation running for three days.

David Sarnoff's lavish spending on NBC's radio studios grated on many of RCA's 288,000 stockholders, whose shriveled shares no longer paid a cash dividend. Even more aggravating, however, was the optimistic executive's unwavering insistence on pouring their money into television research in the middle of the Great Depression. After six years and unknown millions invested in the futuristic technology, television remained many years away from becoming a viable business—as even David Sarnoff freely admitted. His doubters' criticisms were neatly summed up in the sarcastic nickname they invented for him: "televisionary."

THE IDEA OF USING electromagnetic waves to deliver moving pictures had been batted around since Sarnoff's early days at the Marconi Wireless Telegraph Company. Most efforts in 1920 involved rapidly spinning disks inside every television set, a mechanical system that proved hard to manufacture and incapable of producing a decent image on the screen. Sarnoff backed an "all-electronic" version of television that did away with the need for a spinning wheel in every set. That idea had originated in 1908, when a British scientist proposed to use beams of electrons to create television pictures. It took until 1927 for an American inventor named Philo Farnsworth to finally build a basic television set based on the principle. Still, even after that breakthrough, many essential pieces of the TV puzzle remained undiscovered.

While television was often discussed as a single invention, Sarnoff knew that it would need a wide assortment of components to work. Even the wireless telegraph—the first and simplest form of wireless communications—had required the U.S. Navy to force competing inventors into a patent pool before a viable system could be put on the market. AM radio technology was even more complex. RCA now owned more than four thousand patents covering AM radio components. Television promised to multiply that number once again—and in the process ignite a new round of litigation over how to set the license fees paid to inventors and how those fees should be divided up.

To David Sarnoff, the main challenge facing the television

industry was familiar: how to consolidate all the inventions into a single patent pool that manufacturers could license at a reasonable rate. Even Philo Farnsworth's all-electronic TV was useless as a stand-alone product. Among many other challenges, TVs required a camera capable of capturing incoming light waves and inscribing the image they carried onto other electromagnetic waves that could deliver the picture to distant viewers. Sarnoff's top TV scientist, Vladimir Zworykin, had recently made great strides in increasing the light sensitivity of a TV camera, creating a sensor with individual picture elements (pixels, as they would come to be called) that could store the light's energy for a split second before releasing it in a burst. Zworykin's device effectively amplified the light and allowed electronic TV cameras to capture images that looked like the real world for the first time. Still, as Sarnoff freely admitted, even if RCA could combine the best work of Philo Farnsworth, Vladimir Zworykin, Allen DuMont, and every other brilliant engineer who had tackled a piece of the puzzle, the technology to create a viable television industry simply did not yet exist.

And yet David Sarnoff's enthusiasm for the new medium knew no bounds. He had ordered every radio studio at 30 Rock wired with one set of cables designed for radio and a second set for television. One NBC studio featured a giant rotating stage built to work with television cameras that didn't yet exist. "This place has, I can tell you, the fabulous quality," a reporter for the *New Yorker* wrote after visiting Radio City and the free-spending visionary who defied the Depression to build it. "It is a new and higher swell upon the era in which all of us are struggling so bravely, neck deep in the miraculous."

Ignoring the skepticism and scorn of his many critics, Sarnoff continued to proselytize for his vision of a glorious future. "The expansion of the useful radio spectrum has only begun," he had said in a recent speech, summarizing the philosophy he had shared with Edwin Armstrong for the past twenty years. "Science repeatedly has shown its ability to transcend the limitations of the human intellect. It has crashed through physical barriers too vast for our minds to encompass. It has harnessed natural forces that we can hardly define, let alone understand."

WITH DAVID SARNOFF in the passenger seat, Armstrong's sports car weaved through the crowds of last-minute Christmas shoppers, heading uptown toward Columbia University. David Sarnoff didn't share his friend's love of fast cars, which he viewed as part of the unsettling wild streak that Armstrong had demonstrated over the years.

Nearly a decade earlier, after licensing another one of his inventions to RCA in exchange for a small fortune in cash and RCA stock, the inventor had celebrated by breaking into RCA's radio station in midtown Manhattan. He then climbed to the top of the giant radio tower, posed for a picture while balancing on one foot, and sent a copy to Sarnoff. That stunt provoked a brief falling-out after an unamused Sarnoff dashed off a sharply worded letter in response: "If you have made up your mind that this mundane world of ours is not a suitable place for you to be spending your time in, I don't want to quarrel with your decision, but keep away from the Aeolian Hall towers or any other property of the Radio Corporation."

The two friends had put that spat behind them quickly, and in the years that followed, Edwin Armstrong became such a regular visitor at David Sarnoff's Upper East Side town house that Sarnoff's sons, Robert, Edward, and Thomas, nicknamed their father's regular morning companion "the coffee man." Still, Sarnoff's concerns over his friend's behavior had never entirely gone away—and in the last five years they had come back stronger than ever.

Recently, Armstrong had developed an unsettling habit of getting into fights that he didn't need to start, stood little chance of winning, and never seemed to put behind him. Of most concern, he continued to wage a legal battle that Sarnoff considered both absurd and unhealthy. As the twentieth anniversary of their night in Belmar approached, Major Armstrong continued to fight over the rights to his very first discovery. The inventor's interminable patent battle was now before the Supreme Court of the United States, the second time the dispute had reached the nation's highest court and the thirteenth courtroom to hear the case.

Like essentially every new wireless invention, Armstrong's vacuum tube amplifier had been built atop the work of several other inventors. Thomas Edison got things started in 1875, when he noticed peculiar changes in electricity passing through a glass tube from which most of the air had been removed. John Ambrose Fleming, British Marconi's star engineer, made the next leap in 1905, putting a pair of electrodes in the vacuum tube that allowed it to direct the flow of electricity. Lee de Forest, an American inventor, added a small grid in Fleming's tube, which had the surprising result of allowing the tube to ever so slightly amplify an electromagnetic wave's power, for reasons de Forest couldn't explain. Six years later, while still in college, Armstrong set out to determine how de Forest's device worked. Soon he hit upon the idea of intentionally feeding a wireless signal back into itself, an epiphany that led to the thousandfold amplification that had wowed Sarnoff and eventually led to Armstrong's discovery that the device could be turned into a tool for creating the waves as well.

This complicated history, combined with the technical nature of the advances, soon proved too much for a patent system designed to handle the simpler discoveries of an earlier age. In 1914, the U.S. Patent Office declared that Armstrong's amplifier patent had priority over de Forest's invention. Unwilling to accept that his work was merely a stepping-stone to Armstrong's fundamental insight into amplification, de Forest sued. He claimed that his notebooks showed that he had hit upon the same idea first, even though he never patented, publicized, or otherwise made use of the insight until after Armstrong had published a paper describing the inner workings of his device. Their fellow engineers dismissed de Forest's claims as self-serving nonsense, and when the Institute for Radio Engineers voted to give Armstrong its first Medal of Honor for the invention, the issue seemed settled.

And so it would have remained, if Armstrong had not decided to reopen the case by suing de Forest for patent infringement in the months following World War I. That case started well for the Major, who in 1921 won his first court victory when a judge rejected de Forest's claims to have made the discovery first. De Forest had patented even minor, barely useful discoveries, and the judge scoffed at the idea that he would have chosen not to patent the biggest breakthrough in the history of radio.

While de Forest was no match for Armstrong as an inventor, as a legal strategist he proved far superior. Armstrong, by restarting the hostilities, opened the door for his rival to challenge the original verdict of the patent office. This time around, de Forest's lawyers blurred the technical issues so effectively that the patent court declared, for the first time, that it was more likely than not that de Forest had figured out how the device worked in the first place. Endless appeals followed, but now the legal battlefield was tilted in de Forest's favor. Following a legal logic that Armstrong seemed unwilling to acknowledge, the higher courts refused to relitigate the technical findings, instead treating them as an established fact. Finally, in 1928, the Supreme Court ruled against Armstrong, saying it did not have grounds to overturn the lower court's finding. By that point, the major patents being fought over were about to expire, making further appeals pointless, since there was nothing of value left to wrangle over.

And yet Major Armstrong was not done fighting. What he wanted most, Sarnoff knew, was not the money that came from owning a patent but the legal acknowledgment of what he liked to call "engineering reality." What particularly angered him was the scientific ignorance behind the court's ruling and the way de Forest had designed his legal strategy to capitalize on that ignorance. In 1930, almost two decades after he invented the amplifier, Armstrong hit upon creative way to give himself one more chance to beat de Forest in court. He approached a small radio manufacturer being sued for violating one of de Forest's unexpired patents and offered to pay the firm's legal expenses if it agreed to fight the case through to the end.

The Major's unyielding litigation strategy put David Sarnoff in an awkward position. Sarnoff had first encountered

Lee de Forest when the inventor filed a patent claim against the Marconi Wireless Telegraph Company of America. In his experience, de Forest was a vain, mean-spirited, and unprincipled man whose talent as an engineer was no match for Edwin Armstrong. When Armstrong's mentor Michael Pupin died, de Forest had used the opportunity to send a taunting and sarcastic condolence telegram to the grief-striken inventor. De Forest also had a habit of getting involved in stock-selling schemes that had resulted in jail terms for many of his previous business partners. However, since RCA also owned a license to the de Forest patent that Armstrong was fighting to invalidate, Sarnoff's lawyers had been forced to back de Forest's side of the case.

The lawyers advising Sarnoff told him that his friend's latest litigation scheme had left the inventor in a rotten legal position. While Armstrong viewed his legal fight as a battle for justice, the judges hearing the case now saw the relentless engineer as a vexatious litigant. In 1930, Armstrong lost the first round of the case when a federal district court judge ruled against him. Undeterred, he appealed and persuaded a panel of appellate judges to reverse the lower court's ruling. Now the case was back in front of the Supreme Court, where Major Armstrong seemed confident of his chances of winning—far too confident, as far as David Sarnoff could tell.

Sarnoff could never fathom why Armstrong had picked this fight, especially on legal terrain that put him at a huge disadvantage. It pained him to see his friend endure the damage inflicted by his own self-destructive bullheadedness. Sarnoff, by contrast, took a pragmatic and dispassionate view of RCA's involvement in the case. As the manager of a company built around a pool of thousands of patents, he couldn't contradict the courts and unilaterally void a patent owned by his shareholders. A legal fight had to be fought on the legal system's terms, not according to Professor Edwin Armstrong's principles of engineering.

ARMSTRONG FOUND A PARKING SPACE at Columbia University and led Sarnoff across campus to Philosophy Hall, seeming much like the giddy kid who had shown him his first amplifier almost twenty years before. They went downstairs to the basement. Here, in the Marcellus Hartley Laboratory, which Major Armstrong ran, Sarnoff could see a tableful of radio gear laid out in an unusual order.

The two had often discussed ways to solve the problems that had always dogged radio, chief among them the issue of static. Electric razors and other appliances frequently interfered with AM radio signals, warping Louis Armstrong's riffs and muffling Jack Benny's punch lines. Since the days of the Marconi Wireless Telegraph Company, thunderstorms over the Atlantic had blocked intercontinental text messages. Even a car's sparkplug, which emitted far less energy than a lightning bolt, could scramble reception in nearby radios.

The Major, rarely happier than when revealing a great discovery, began to set up his demonstration. Years ago David Sarnoff had coined a nickname for the elusive device that could solve that static problem, dubbing the as-yet-mythological gadget "the little black box." Pointing to one of the two circuits laid out on the table in front of Sarnoff, which looked perfectly ordinary, the Major said it was a typical, static-prone AM receiver. The other, he declared proudly, was hooked to the little black box.

Armstrong flipped on the second receiver. Suddenly the faint buzz of static coming out of the speaker dropped away, while the music itself gained a fuller, richer sound. Sarnoff employed many people with expertise in sound quality, but he didn't need their precise tests to tell him what his ear made clear. The music was coming through with almost lifelike clarity, especially the higher notes and the "overtones" that got lost in standard AM radio transmissions.

While the fidelity was remarkable, it was hard for the RCA president to say how much of of it was due to Armstrong's new invention and how much was the result of his top-of-the-line gear. The demonstration certainly fell short of the jaw-dropping results Sarnoff remembered so well from their shared night in Belmar.

To people familiar only with the basics, the difference between FM and AM radio seemed inconsequential. All radio works by imprinting a sound wave onto an electromagnetic wave that carries it to its destination, then reversing the process to turn the incoming airwave back into sound. Traditionally, radio transmitters imprinted the sound onto the outgoing electromagnetic wave by varying the wave's power level, and thus its height, or amplitude. Armstrong's new system did it by varying the length of the electromagnetic waves instead. (In engineer-speak, AM radios were said to modulate an airwaves' amplitude while FM radios modulated their frequency, *frequency* being another word engineers use to describe a wave's length.)

At first blush, transmitting waves of varying heights or transmitting waves of varying lengths seemed like a wash, a distinction without a difference. The Major tried to explain to

Sarnoff why he believed the new system would be so important. Ever since the days of Hertz, people had been working on better ways to imprint information into electromagnetic waves. AM radio engineers attacked the problem by attempting to come as close as possible to emitting a single electromagnetic wave with a fixed length. Now Armstrong argued that an even better system would use a large band of wavelengths at the same time, a new style of "broadband" communications that would mark the start of a new era, in which communications networks would be able to carry far more information than older, "narrowband systems" such as AM radio.

If he hadn't known Armstrong so well, David Sarnoff might have dismissed the idea out of hand. According to engineers at RCA, AT&T, and throughout academia, the question of how best to transmit information using electromagnetic waves had long been settled—and it was the exact opposite of what Armstrong now claimed. While the idea of imprinting information onto waves of varying length had occurred to many early inventors, it was never worth pursuing. The Danish inventor Valdemar Poulsen, who built the first transmitter cable that could pull off the trick, declared in 1906: "The process of altering the length of the emitted wave must be abandoned fundamentally, since this implies that each sending station would be characterized by two waves, and thus the number of stations which can work on the same service would be reduced to one half." In 1922, an AT&T researcher had published a paper in the prestigious *Proceedings of the Institute of Radio Engineers* that claimed to prove mathematically that frequency modulation "inherently distorts without any compensating advantages whatsoever."

Ever since, Major Armstrong told Sarnoff with a grin, experts around the world had dismissed the idea: "The textbooks testify unanimously to the superiority of amplitude modulation."

Though he lacked Armstrong's technical acumen, Sarnoff was good at detecting the fundamental differences between the two competing technologies. He appreciated the basic scientific advance in Armstrong's FM system, but he found himself conflicted—or at least confused—about the importance of the discovery. Part of his uncertainty, he told Armstrong, came from having seen a demonstration only in a controlled laboratory setting, knowing that tests out in the real world often turned up unforeseen problems.

The fact that a technology was "technically possible" and "greatly superior" was not the end of the conversation. Had Armstrong considered that converting the radio industry to FM would mean junking tens of millions of radios and millions of dollars more in radio station transmitters? Had he considered the difficulty—perhaps even the impossibility—of creating an FM network, since the AT&T wires that carried NBC's AM radio programs around the country were incapable of passing on the richer tones of FM?

For his part, David Sarnoff wasn't ready to treat FM radio with the same degree of passionate enthusiasm he had given Armstrong's earlier discoveries, nor with the passion he now devoted to TV. Sarnoff envisioned television as the new market, the new adventure, the next great wireless app. Still, he was happy for his friend. Once again, Major Edwin Armstrong was rewriting the radio engineering textbooks. "This is no ordinary technology," Sarnoff told Armstrong as they wrapped up their evening in the laboratory. "This is a revolution."

Leah Sarnoff with five-year-old David (left) and his younger brothers, Morris and Lew, in 1896. Four years later they would flee their village in the Russian Empire and sail for New York. (Courtesy of the Hagley Museum & Library)

In 1906, fifteen-year-old David Sarnoff landed a job as a messenger boy with the world's first wireless communications company. The following year, when this picture was taken, he had become the Marconi Wireless Telegraph Company's youngest operator, spending his days sending Morse code messages to passing steamships. (Courtesy of the Hagley Museum & Library)

Guglielmo Marconi, on a 1915 voyage to New York. The Italian won fame as an inventor, though the success of his pioneering wireless company owed more to his flair for self-promotion, his talent for opening investors' wallets, and an unshakable faith in the power of the airwaves. (Courtesy of the Hagley Museum & Library)

A group of famous American scientists at RCA's new transatlantic wireless station in 1921: Pioneering electrical engineer Charles Proteus Steinmetz (the short man in the light suit), Albert Einstein (standing to the right of Steinmetz), and future Nobel laureate Irving Langmuir (standing to the left of Steinmetz). RCA's commercial manager David Sarnoff (on the left, leaning in) led the tour.

Though Guglielmo Marconi claimed to have sent a text message across the Atlantic Ocean in 1901, it was not until two decades later, when this station opened in New Brunswick, New Jersey, that cheap waves zooming over the Atlantic began challenge the web of expensive telegraph cables running beneath the ocean. (RCA)

In 1930, Edwin Armstrong (right) tracked down this dilapidated shack, which the American Marconi Company had used twenty-eight years earlier to house one of its first wireless telegraph stations. In a sentimental gesture, Armstrong bought the shack and shipped the building to David Sarnoff as a gift. When Guglielmo Marconi (left) visited the United States in 1933, the two friends took him to see it. (Courtesy of the Hagley Museum & Library)

As a young executive at the Radio Corporation of America in the 1920s, David Sarnoff pushed the company to shift into a new business: radio broadcasting. Still, the former wireless operator always retained an affection for the first wireless app—sending text messages using Morse code—and kept a telegraph key at his desk to trade text messages with RCA wireless stations around the globe. (RCA)

Commerce Secretary Herbert Hoover (far right) convened a conference in 1924 to debate which laws should rule the airwaves. Sarnoff (second from left) quickly made enemies, terrifying local radio stations and newspapers with his plan to "vastly increase" the number of radio stations on the AM dial. (The Library of Congress)

RCA and NBC's new home at "30 Rock" included lavish studios for NBC radio performers like Eddie Cantor and Jack Pearl. New York's largest theater, Radio City Music Hall, was located at ground level, while David Sarnoff ruled his empire from an office on the fifty-third floor. (Via Pixabay; RCA/NBC)

NBC
NBC
NBC

In 1933, the doggedly optimistic David Sarnoff risked his career by moving RCA and NBC into Rockefeller Center. Brushing off the ongoing Great Depression, Sarnoff signed the largest commercial lease in New York history. (Photograph by Samuel H Gottscho, 1933)

David Sarnoff makes the first television broadcast to the general public at the 1939 World's Fair. While almost no one saw the speech, since almost no one in 1939 owned a TV set, the fair began Sarnoff's effort to convince the public that buying a set would soon make sense—and that he had been right to spend the past decade pouring RCA's research budget into the unproven technology. (RCA/NBC)

By the time radio broadcasting took off in the 1920s, thousands of useful wireless components had won patents—but one invention stood apart from the rest: the vacuum tube. Thomas Edison first noticed electric currents behaving oddly inside glass tubes emptied of air. John Fleming added additional circuitry inside a vacuum tube that gave it the power to direct the flow of electricity. Then, in 1914, Edwin Armstrong made the biggest leap of all, building a new type of vacuum tube around three key circuits—a device called a triode vacuum tube, which was able to create and greatly amplify electromagnetic waves.

Increasingly complex vacuum tubes formed the heart of virtually all electronic equipment for the next forty-four years—until two Bell Labs scientists unveiled the device that would finally make Armstrong's invention obsolete by coaxing the same tricks out of tiny solid chunks of silicon and other semiconductors. Their paper, "The Transistor, A Semi-Conductor Triode," would serve as the obituary for the vacuum tube age. (Courtesy of the Hagley Museum & Library)

Lizzete Sarnoff, like her husband and three sons, volunteered for military service during World War II, in her case as a nurse. (American Red Cross)

David Sarnoff as a colonel in the U.S. Army Signal Corps, before his promotion to brigadier general. (Photo by U.S. Army Signal Corps)

By the 1950s, as the space age began, Sarnoff became entranced with the idea of using satellites to leap past AT&T's network of telephone wires. (Courtesy of the Hagley Museum & Library)

Sarnoff with AT&T chief executive Frederick Kappel (above, right) in 1966. Executives at the phone monopoly first grew to dislike Sarnoff in 1923, beginning a bitter rivalry that would span four decades. (RCA)

David Sarnoff with Vice President Lyndon Johnson in 1961. LBJ used his influence at the major networks to help his wife's media empire, developing particularly close ties with NBC's main rival, CBS. (Courtesy of the Hagley Museum & Library)

At the end of David Sarnoff's career, a new threat to his legacy at RCA and NBC emerged: the incompetence of his son and chosen successor, Robert (standing, right). (RCA)

David Sarnoff with CBS chief Bill Paley (left) and ABC's Leonard Goldenson (right). The trio controlled "The Big Three" television and radio networks, which earned them prestige and power—but very little profit. The real money in broadcasting wound up in the pockets of local TV stations. (RCA)

Sarnoff at the Waldorf Astoria in 1966, commemorating a career in communications that stretched from the early days of the wireless telegraph to the dawn of the Internet era. The dots and dashes on the podium spell out "60 years." (Courtesy of the Hagley Museum & Library)

AFTER THE HOLIDAYS, Sarnoff packed up his family for a vacation in Bermuda while Armstrong began moving his FM gear into the RCA labs on the eighty-eighth floor of the Empire State Building, which Sarnoff had made available for his use. At the end of January, the inventor interrupted his work to cable Sarnoff in Bermuda, commemorating the twentieth anniversary of "that cold night in Belmar" when the two young believers shared their first glimpse of the limitless powers of the air. In his heartfelt reply, David Sarnoff made no secret of his affection for his old friend, or his concern.

> *"Dear Major,*
>
> *"It was very kind and thoughtful of you to have sent me your message to Bermuda, which brightened my vacation by making me think more of the happy past than of the confused present.*
>
> *"Well do I remember that memorable night at the Belmar Station when, by means of your 'magic box,' I was able to copy the signals from Honolulu and other distant parts of the world. The floors at Belmar did not then seem to feel so hard and the bitter cold did not numb the fingers. Whatever chills the air produced were more than extinguished by the warmth of the thrill which came to me at hearing for the first time signals from across the Atlantic and across the Pacific.*
>
> *"It would seem hard to believe that twenty years—a full generation—have passed since the night of the event we now speak of. One's natural vanity would prefer to explain this by emphasizing the youth of the wireless art, but the generous sprinkling of grey hair that now resides on my dome and the pronounced lack of grey or black hair on your dome, alas, make it impossible for you and me to embrace that alibi.*

"And yet here you are a generation after that event, still gripped by the mystery of the air, still challenged by the secrets of space, and still in the forefront of advanced thinkers and workers in the art. These attributes are true evidence of youth, the others but fleeting signs.

"It is not a sermon I wish to preach, but rather a hope to emphasize. If you will fix your gaze and energies upon the next twenty years and let history deal with the past twenty years, the telegrams and letters we should be able to exchange at the end of the next generation would make us feel that we were still young even then.

With affectionate regards,
Sincerely yours,
David Sarnoff"

TWO MONTHS LATER, as Sarnoff had expected, the Supreme Court shot down Armstrong's long-shot legal challenge. The news interrupted Armstrong's FM work and sent him spiraling into depression. Instead of seeing the case as Armstrong did—as an opportunity to acknowledge and reverse a scientific error—the Supreme Court used it to discourage other wealthy litigants from following in his footsteps. Justice Cardozo's opinion, which reversed the appeals court and rejected Armstrong's claims, focused not on the technical merits of the patents but on the need for tighter standards to prevent other patent cases from dragging on for two decades.

The court's decision went out of its way to make clear that the ruling was not a verdict on the engineering debate at the heart of the case. The court clearly doubted de Forest's complicated justifications for why, even though he claimed to have made the important discovery before Armstrong, he didn't file for a patent until after he learned how Armstrong's amplifier worked. Regardless, Cardozo wrote, "These explanations, even if not wholly convincing, are not so manifestly inadequate as to lead us to say that the [invention] has been proved in any clear or certain way to have been developed and applied by Armstrong before it was born in de Forest's mind."

That logic infuriated Armstrong. To him the evidence was so obvious, the engineering mistakes of the patent courts so glaring, that he couldn't conceive of how any fair-minded person could see it differently.

A few weeks after the verdict was announced, Sarnoff watched his embittered friend attempt to return his Medal of Honor to the Institute of Radio Engineers. The inventor explained that his fellow engineers had awarded him the medal for an invention that the highest court in the land now said he hadn't made. The institute's board, including many RCA employees, insisted that Armstrong deserved the award and should keep it. The meeting ended as the emotional inventor thanked his colleagues and described his legal odyssey this way: "It is a long time since I have attended a gathering of the scientific and engineering world, a world in which I am at home, one in which men deal with realities and where truth is, in fact, the goal. I have been an exile from this world and an explorer in another, a world where men substitute words for realities and then talk about the words."

In the months following their meeting at Columbia, the two friends began to drift apart as they pursued their separate interests. In late 1934, Sarnoff asked Armstrong to clear out his FM radio equipment from the Empire State Building lab so RCA could begin the television test it had originally designed the lab to conduct. (Broadcasting either FM or TV signals from atop the world's tallest building offered several unique experimental opportunities.) Armstrong, who thought TV technology was not yet ready for full field tests, pushed back, sparking a heated argument between the two men.

Still, a deep affection remained. At the 1935 RCA shareholder meeting, after other RCA shareholders again attacked David Sarnoff for pouring millions into television research and other spending, Major Armstrong rose to defend him. Anyone who doubted his old friend's technical acumen or vision should have his head examined, the inventor declared. Armstrong acknowledged that he personally didn't share Sarnoff's rabid enthusiasm for TV and had been squabbling with him over whether to push FM radio or television to market first. Still, he cautioned RCA's other shareholders: "I think you would have been wiped out if it hadn't been for him. I tell you, I wouldn't have his job for $500,000 a year.

"We don't agree on everything," Armstrong added. "I have a row on with him now. I am going to fight it through to the last ditch. I just thought you should know what you owe him."

To the last ditch. That was his friend's style of fighting, all right. Still, the inventor's words touched Sarnoff.

He sent a letter to Armstrong a few days later, seeing a chance to repair their rift. "Doubtless I have made many mistakes in my life," David Sarnoff wrote, "but I am glad to say they have not been in the quality of the friends I selected for reposing my faith."

CHAPTER 10

ARMSTRONG

Washington, D.C.—1945

MAJOR EDWIN HOWARD ARMSTRONG DISLIKED WASHINGTON, D.C., and the muggy June air promised a miserable day in the offices of the Federal Communications Commission. Still, he had learned that visits to the capital were often necessary to head off the sort of muddleheaded laws and regulations that could cripple new technologies. This was the fourth hearing he'd been forced to attend just to quash one crackpot plan that would turn $75 million in consumer FM radios into obsolete junk.

Though he still bore the scars of his long fight with de Forest, the glum Major Armstrong of 1934 was gone, replaced by an inventor who once again fixed his gaze firmly on the future. Rather than relitigating old lawsuits, Armstrong now split his time between getting the fledgling civilian FM radio industry off the ground and wrapping up his wartime work for the Army Signal Corps. He'd spent much of the previous few months on a new type of radar. He planned to demonstrate that invention in a few months by bouncing a radio wave off the surface of the moon, a feat no other radar system could come close to matching.

Major Armstrong's career and spirit had been lifted by the remarkable performance of FM in the dozen years since he first showed David Sarnoff the technology. Sarnoff's offer to let him test his FM transmitters in the RCA's Empire State Building laboratory helped get the technology off to a fast start, as did the platoon of RCA's best young engineers that Sarnoff sent to assist him. Within four months, FM signals were blasting out from New York's tallest building, and the reports that Armstrong and his RCA assistants filed on the signal quality reaching Long Island, New Jersey, and Connecticut ranged from encouraging to astounding.

In one experiment, an FM signal sent with two kilowatts of power delivered better sound than an AM radio station using fifty kilowatts. A giddy Armstrong wrote in his lab notes that the tests "surpassed all expectations" and bragged that "the margin of superiority of the frequency modulation system over amplitude modulation was so great that it was at once obvious that comparisons of the two were principally of academic interest."

The new technology also revealed unexpected capabilities that had nothing to do with Armstrong's original goal of making a high-fidelity, static-resistant radio service. FM transmission worked on wavelengths too short for AM radio to handle, meaning FM opened up the possibility of adding thousands more radio stations to the dial.

FM radio also revealed an unexpected economic advantage.

Since David Sarnoff negotiated RCA's cease-fire with AT&T in 1926, the radio networks had been the phone company's largest customers, dependent on its network of telephone wires to deliver the sound of Jack Benny, Minnie Pearl, Frank Sinatra, and all the other radio stars to local broadcasting stations around the country. NBC was its bitter rival's single largest customer, paying more than $3 million per year to lease the phone lines that moved its radio programs around the country. In the AM world, hitching a ride on telephone lines made sense, since rebroadcasting an AM radio signal injected unacceptable amounts of static into the copy. FM signals, on the other hand, could be picked up and rebroadcast—meaning they could simply hop over AT&T's expensive wires.

To help jump-start the industry in the late 1930s, the Major dipped into the fortune he had amassed by licensing his earlier inventions. On the Hudson River palisades, he built a giant antenna designed to blanket New York City (along with much of Connecticut and New Jersey) with an FM signal. He underwrote the creation of an FM network in order to show off the technology's ability to link distant stations without using expensive telephone wires. The Federal Communications Commission had been slow to approve many of Armstrong's experiments, but the technology's advantages were too overwhelming to ignore. In 1940, the commissioners agreed to allocate a range of wavelengths, between five and six meters, for FM radios.

If FM's takeover of the commercial radio market had seemed likely before the Second World War, its performance on the battlefield eliminated any remaining doubt. At the start of the conflict, Major Armstrong had pushed every branch of the military to quickly switch to his technology, waiving any licensing fees he might be owed, in order to lobby for the rapid adoption of FM in the military without being accused of trying to line his own pockets.

The military poured $100 million into FM equipment during the war, and generals and admirals alike testified to the advantages it offered. In Europe, FM gave General Patton's tanks a key edge over the AM-equipped panzers that often had to stop moving in order to communicate with each other. (Otherwise the low-fidelity AM signal was drowned out by the roar of the engine and its accompanying static.) In the Pacific, troops used "walkie-talkies," portable FM radios that gave island-hopping marines a critical edge in battlefield communications.

AS THE WAR WOUND DOWN, the Federal Communications Commission began preparing for the civilian industries that would share the use of the airwaves after the war—most notably AM radio, FM radio, television, and radar. In 1944, the chairman of the FCC, Lawrence Fly, convened a "Radio and Television Planning Board" and recruited the nation's top engineers to join it. In proceedings that impressed Major Armstrong with their thoroughness and engineering rigor, he and twenty-seven other experts spent several months studying and debating which airwaves should be set aside for the infant FM industry, along with a host of related technical issues.

Four days before the panel of experts was set to conclude their year-long study, however, a radio engineer named Kenneth Norton showed up carrying binders full of charts and equations that, according to Norton, revealed a hidden danger to the civilian FM industry. While the new technology had worked well so far, he said, an increase in sunspots would make FM reception go haywire in the coming years—unless the FCC took radical action to head off the threat. Norton recommended switching FM radio stations onto much shorter airwaves that would make all existing FM radios useless but which, his theory predicted, would be safe from the sunspot interference.

To Armstrong and his fellow radio propagation experts on the panel, Norton's sunspot theory had a superficial plausibility. As Guglielmo Marconi discovered when he tried to send daytime messages across the Atlantic, the sun's rays often altered the atmosphere's ability to bounce radio waves back to earth. Astronomers knew that sunspots (and the bursts of solar radiation they release) tended to ebb and flow in an eleven-year cycle, a cycle that would be peaking during the next few years. Beyond that, however, Armstrong and his colleagues quickly determined that everything else about Norton's theoretical threat was bunk. They voted twenty-seven to one in favor of keeping FM in its prewar spot on the dial and ignoring Norton altogether. (The lone "no" vote came from an engineer at DuMont Laboratories, a television company that wanted FM's airwaves reserved for that industry.)

A rotund man with horn-rimmed glasses and a haughty air, Norton embodied everything Major Armstrong disdained in an engineer, "Mr. Norton is a theoretician—without practical experience in the art, and with implicit confidence in his ability to sit in back of a desk and by means of paper and pencil and a slide rule to work out the answer to any problem," Armstrong scoffed. "Now, those of us who have been in practical engineering for 30 years know that that just cannot be done. We have been fooled too many times by our theoretical calculations."

After the FCC panel of experts finished their study, which had included an exhaustive look at the effects of atmospheric reflectivity, Armstrong assumed that the threat of Norton's pseudoscience had been quashed. Despite the experts' consensus, the newly appointed FCC chairman who took over following Chairman Fly's surprise resignation, continued to express concern over the sunspot threat. At his next meeting with the FCC, Armstrong and an array of other top scientists made sure to detail Norton's many errors for Paul Porter, the new chairman, and his colleagues. The head of MIT's Laboratory for Cosmic Terrestrial Research took the platform to point out that Norton's recommendation was based on a coming outbreak of sunspots that might never occur, that no reception data backing Norton's theory existed, and that the data that did exist contradicted his theory.

Even if Norton was right, the MIT physicist added, FM would still provide a far more reliable service to the country than AM radio was currently doing, thanks to its resistance to the thunderstorms and man-made sources of interference that regularly drowned out AM radio programs. Zenith's vice president of engineering presented extensive experimental data that directly contradicted Norton's theories. Karl Jansky, the head of the FM radio panel, and himself famous for discovering cosmic radiation while working at Bell Labs, summarized the scientists' case in simple terms:

> "From a propagation standpoint the issue is clear. Either you believe Norton in spite of the errors he has made and

> the unproven assumptions he has had to adopt to support his conclusion that FM should be moved upward in the spectrum, or you believe Dellinger, Beverage, Burrows, Stetson, Pickard, Bailey and Armstrong.
>
> "Unproven theory says one thing. Practice and experience and the expert opinions of others say another."

Jansky concluded by quoting Leonardo da Vinci: "'The supreme misfortune is when theory outstrips performance.' I know of no better characterization of the situation which confronts this Commission with respect to the formation of a decision upon what is obviously one of the most complex fields of science but a decision which, gentlemen, is of the utmost importance to the future of the entire radio industry."

After that hearing, Armstrong again assumed that this would finally put the sunspot matter to rest—and again he was quickly proven wrong. In what appeared to be a face-saving bluff, Norton reappeared with a new story: he had found secret military data on FM radio propagation, and that data verified his sunspot theories. Armstrong, Jansky, and several other members of the panel who had their own military security clearances demanded that Norton share his secret data, and the military arranged a special classified hearing.

At that hearing, Armstrong and his allies exposed major flaws in Norton's logic and arithmetic yet again. In one particularly glaring error, Norton predicted that sunspots would cause FM radio signals from South America to bounce off the "F2 layer" of the ionosphere and land in the United States. Armstrong pointed out that this theory violated not just the laws of physics but also those of geometry—since the height of the

F2 layer made it impossible for an airwave to reach the United States by banking off that part of the atmosphere a single time, as Norton's equations assumed.

Strangely, even after the experts demolished his arguments, the threat posed by Norton *still* didn't go away. In the unclassified synopsis of Norton's classified testimony, Major Armstrong had been shocked to find a summary that changed the wording of the classified report. In response to objections, the report now said that Norton had furnished "a satisfactory explanation of his theories."

The primary purpose of Armstrong's latest trip to Washington was to make sure the error in the public report was eliminated and Norton's kooky theory dispensed with before it derailed the new industry. It was an inconvenience to Armstrong to have to spend so much of his time keeping the country's technology policy on track. Nonetheless, he was pleased that his attentiveness ensured that he would be able to stamp out bogus, self-serving science before it did any permanent damage.

PAUL PORTER, the FCC's new chairman, gaveled the hearing to order. A large man with a gravelly voice and a talent for backslapping, Porter had made a career of jumping between Washington's countless obscure bureaucracies, devoted to fixing flaws in the market. At the Agricultural Adjustment Administration this involved reducing crop surpluses by paying farmers to leave fields fallow. As chief of the rent control section of the Office of Price Administration, Porter told landlords what they could charge tenants. Most recently he had worked for CBS as the network's Washington counsel.

While Kenneth Norton personified everything Armstrong disliked about Washington engineers, Paul Porter embodied everything he found disagreeable in lawyers. Under Porter's direction, the day's hearing dragged on, and as the clock neared six, Armstrong still hadn't gotten his say. The FCC chairman seemed uninterested in the technical details disproving Norton's claims, appearing almost eager to shut down the hearing without letting Major Armstrong expose the curious alteration in the unclassified version of Norton's testimony.

As the long meeting was wrapping up, Major Armstrong stood to make sure he got his turn to speak. Directing Porter's attention to the unclassified record, he said:

"Now, on page 60 there is a statement at the top of the page with which I cannot agree: *'A satisfactory explanation regarding the appropriate method to be employed in the analysis of the problem was furnished by Mr. Norton during the closed hearing. This analysis indicated that no error had been made in this report.'* Now, that refers to the memoranda prepared by myself, Dr. Beverage, and Dr. Burrows."

Porter cut Armstrong off with a meandering objection: "I think it is in the nature of classified testimony as to Mr. Norton's explanation and the letter he put in the record today, and if you desire to submit a further statement in writing, based upon Mr. Norton's letter, which is available to you, you may do so, but I do not think it is profitable to discuss it further because the record is very clear as to your analysis [and] to . . . Mr. Norton's response."

Armstrong bulled ahead: "The confidential report admits the error. This one denies it. The public report denies the error."

An FCC lawyer jumped in: "Major Armstrong, irrespective of that, do you agree with the facts as stated in the confidential report?"

"If you interpret them as I do, meaning that in the confidential report the error is admitted, then I agree with you."

"I do not think it would be profitable to pursue this matter," Porter said again. "It is getting into a question of semantics and forensics."

"Well, Mr. Chairman, may I say that the public reputation of six men is at stake here. This controversy has been watched throughout the engineering world, the technical world," Armstrong said, pausing briefly to distinguish the world he loved from the world he disdained. "The Commission's public report says that we were wrong. We cannot let that rest before the technical world. We cannot do that."

"Well, make whatever statement you want to about it in the record. You object to the use of the word 'satisfactory,' I take it?"

"Yes, and any other statement that there was no error committed. The error was committed and it is admitted in the confidential report, and that is all I have to say."

"Well, I think the Commission is well aware of the facts in that situation, Major Armstrong, and what we want is comments upon the conclusions and facts we set in this report with respect to F-2. Will you try to limit it . . . your discussion . . . to that question?"

"I have finished my discussion of F-2, Mr. Chairman, pointing out that these theoretical predictions of the inverse distance happen as shown in practice. And I do hope ...aid some basis for an understanding, or proper appraisal, of how much weight ought to be given to theoretical calculations."

That was the point Edwin Armstrong had come to make. If it took a day of sweating in the Washington heat to make sure good engineering won the day, then he considered that a valuable day's work.

"I assume this concludes the hearing," said Chairman Porter, with apparent relief. "I would like the record to show this, we are deeply grateful, not only to Major Armstrong for his patient and persistent efforts to be helpful to the Commission, but likewise to the members of the RTPB panels and others who have participated in this hearing. It is a difficult question. It is one the Commission will consider hopefully—and I might say, prayerfully—and I am quite certain that whatever the result is, whatever our ultimate decision is, that those of you who have participated in these proceedings will know that it is based upon what our conception is of the public view in the matter and an appraisal of the evidence."

THREE DAYS AFTER Chairman Paul Porter thanked Major Armstrong for helping straighten out the commission on the science of FM radio waves, the commission unanimously voted to heed Norton's advice to shift FM to the shorter wavelengths. From now on, FM stations would broadcast on waves between 2.8 meters and 3.4 meters long, meaning the new FM dial would run from 88 megahertz to 108. Owners of radios designed to tune in waves between 5 and 6 meters in length (with dials running from 42 to 50 [the old band]) would soon have nothing left to tune into.

The purpose of the move, the commission explained, was to give the new technology the best chance of success by protecting it from sunspot-induced interference. Heedless of the fact that twenty-seven of its own experts considered that threat fanciful, Chairman Porter publicly congratulated the commission for the thoroughness and scientific rigor with which it had addressed the problem. Overnight, the FCC's decision made $75 million worth of FM radios obsolete and turned hundreds of thousands of enthusiastic early adopters into embittered ex-FM listeners.

Despite the shocking setback, Armstrong and his allies continued to fight for the new technology, arguing that FM possessed such vast technical and economic advantages that the band shift would slow but not stop its advance. Even after the shift, a total of 429 applicants asked the FCC for the right to build a new FM station. "As a vastly improved method of reception, providing for higher quality of reproduction and freedom from static, FM radio should go ahead very fast," wrote columnist Jack Gould, echoing the industry's optimism. No one professed greater enthusiasm for FM than Chairman Porter, who predicted huge sales of the new FM radio sets.

Then came the next round of surprises from the communications commission. Having protected the FM industry from the hypothetical sunspot threat by changing the FM band, Chairman Porter and his colleagues decided it needed to be protected from another potential danger: monopolies. Given the vast superiority of FM to AM, the commission declared, it made sense to pass preemptive restrictions to assure that the new FM industry did not become dominated by monopolies. At the suggestion of CBS lobbyists who a few months earlier had been his coworkers, Porter pushed through strict limits on the range of FM broadcasts as part of a "single market plan" designed to limit new stations to serving individual cities and towns.

Overnight, a typical FM station that had been able reach listeners 150 miles away saw its range cut 66 percent, to 50 miles and its coverage area drop a whopping 90 percent. Armstrong's own FM station was hit the hardest. The inventor had located his pioneering FM station seventeen miles outside of New York City on the Hudson River palisades, a spot chosen for the natural advantage that height offered in sending signals long distance. Under the new rules, the station's elevation was treated as a disadvantage, forcing a 97 percent cut in power, from 50 kilowatts to 1.2 kilowatts. FCC commissioner George Sterling explained the rationale this way: "To permit higher powered stations in the highly congested portions of the United States would of course greatly reduce the number of assignments that could be made in such areas and would thus tend to foster monopoly in FM broadcasting."

Fear of FM monopolies also served as the commission's rationale for denying the technology its two other great advantages: better economics and better sound quality. As part of its single-market plan, the FCC effectively eliminated FM stations' ability to send programs to neighboring markets through the air, forcing them to use AT&T's expensive and low-fidelity telephone wires.

The flurry of restrictions on FM was not designed to shield the profits of the existing AM radio stations, Chairman Porter and his colleagues assured the public, but were simply meant to help FM thrive in the long term. Not only did Armstrong and his allies fail to appreciate the long-term benefits of the band shift and the small-market rules, Porter said, they also greatly exaggerated the costs associated with the new regulations. Rather being forced to junk $75 million in existing FM sets, the FCC chairman promised, owners of existing FM sets would be able to buy ten-dollar converters. Manufacturing new FM radio sets capable of handling the shorter waves would also be a snap. Immediately after the band change, Chairman Porter announced that he had spoken with radio manufacturers who assured him they would have FM radios tuned to the new band within months. As for test data presented by Zenith that purported to show that the band shift would "hopelessly cripple" FM radio, the FCC said its engineers' tests showed the "exact opposite" to be true.

None of it, however, proved to be so. In an attempt to verify Chairman Porter's claim that all major radio manufacturers were ready to start making the new FM sets immediately, *Radio Digest* did its own phone survey of manufacturers—and found that not one was prepared to produce new FM sets. In late 1946, a full year after Porter promised the new sets would be available, the FCC commissioners responded to a question from Congress with professions of surprise that nine out of ten big radio manufacturers still weren't building the new FM sets. The ten-dollar converter also proved to be a figment of Porter's imagination. (This didn't surprise radio experts, since such a device would have to include a shortwave receiver, which could not be built for much less than the price of an entirely new radio.) When the sunspot cycle reached its peak, it brought with it a level of solar activity not seen in nearly two centuries—and even so FM signals did not start pinging around the globe in the way Kenneth Norton had predicted.

As evidence of its errors began to pile up, the FCC casually rebuffed Armstrong, Zenith, and other FM advocates seeking to remedy its mistakes. With the FCC restrictions strangling the fledgling FM industry, forcing many stations into bankruptcy, Armstrong begged the commission to allow long-distance relays of radio shows, which would bolster the industry's finances as well as the variety and sound quality of its broadcasts. Chairman Porter met with Armstrong and his ally, the head of Zenith, helpfully suggesting they apply to get FM's old band of wavelengths returned to the industry so it could use them to establish those relays. They did apply—and eventually Porter decided it wouldn't be a good idea after all: "Upon analyzing the testimony and data, the Commission concluded that the use of the two bands for FM broadcasting was undesirable, that interference would continue to be a serious problem in the old band, and that an excellent FM service would be provided in the high band."

In 1947, Armstrong got another chance to confront Norton at an FCC hearing. After noting that the record-setting flurry of sunspots hadn't caused the predicted problems, Armstrong was shocked to find Norton happy to admit the theory that had crippled the nascent FM industry had been a mistake after all.

"You were wrong?" demanded Armstrong.

"Oh, certainly," Norton replied casually. "I think that can happen frequently to people who make predictions on the basis of partial information. It happens every day."

CHAPTER 11

McCORMACK

New York City—1953

EVEN IN A BUILDING FILLED WITH HIGH-POWERED LAWYERS, Edwin Armstrong's elite legal team stood out as they marched through the lobby, a walking testament to their client's determination to win the antitrust lawsuit he had filed against RCA and expose David Sarnoff as the man who had organized the conspiracy to cripple FM radio. Patent expert Richard Byerly was a founding partner of Byerly, Townsend & Watson and the president of the New York Patent Lawyers Association. William Prickett, from the Delaware-based law firm Prickett, Jones & Elliott, specialized in trying cases before that state's federal circuit court, which was hearing the case of *Edwin H. Armstrong v. Radio Corporation of America and National Broadcasting Company*. The youngest member of the group, Dana Raymond, was a rising star at Cravath, Swaine & Moore and possessed an encyclopedic knowledge of the now five-year-old lawsuit's countless details.

And then there was Alfred McCormack, Major Armstrong's lead attorney, whom many of his colleagues at Cravath considered the most fearsome legal mind in Manhattan. Colonel McCormack had shot to fame during the war. Assigned the task of analyzing the intelligence failures that led to Pearl Harbor, he wound up pushing for a radical restructuring of all military intelligence. McCormack argued for the creation of a single centralized group within the military that would take responsibility for gathering, analyzing, and sharing military information among all branches of the United States armed forces—an argument he made so persuasively that the military brass created the Military Intelligence Service and then appointed him to lead it.

After the war, McCormack had seemed like a lock to become the first director of the Central Intelligence Agency, the new civilian agency he helped President Truman create as a central clearinghouse for all the government's foreign intelligence. After a bitter fight with J. Edgar Hoover, who wanted to keep the job for his own FBI, McCormack decided to take a break from Washington and return to New York for a more lucrative and less pressure-filled job as a lawyer at Cravath, Swaine & Moore, one of New York's elite law firms.

Among McCormack's many talents, his understanding of engineering and science vastly exceeded that of most lawyers and even put many radio engineers to shame. The lawyer had no trouble following the technical side of his client's complaint, and often bandied about terms like *"second harmonic superheterodyne"* to the befuddlement of other lawyers working the case.

McCormack could have easily handled the day's interrogation of David Sarnoff alone, but his wealthy client had developed a wariness that bordered on paranoia when it came to the legal system.

Today was a moment that Major Edwin Armstrong had been anticipating for a long time. Sarnoff's lawyers had subjected him to twenty-five days of depositions. Now it was Sarnoff's turn to testify. In a few minutes, McCormack and his team would get to place David Sarnoff under oath and interrogate him about Armstrong's allegations that he had intentionally crippled the radio and television industries to protect his own empire.

A little before ten thirty in the morning, Armstrong, McCormack, and the other three lawyers walked into the lobby of Cahill, Gordon, Zachry & Reindel, one of the outside law firms that RCA kept on retainer. John Cahill, the firm's founder, welcomed the inventor and his lawyers to his offices, inviting them to sit down in a conference room where the deposition of David Sarnoff was set to occur.

McCormack viewed Cahill with a mixture of respect and wariness. RCA's lead litigator had his own impressive pedigree, having made his name as the U.S. attorney for the Southern District of New York, where he had put a slew of mobsters and even a corrupt federal judge behind bars. Then he had gone to Wall Street, building Cahill, Gordon into one of the financial industry's top litigation firms. Sarnoff clearly trusted Cahill, who had argued several of RCA's cases before the Supreme Court.

It had been a long time since Major Armstrong and General Sarnoff had seen each other. At sixty-two David Sarnoff had become the picture of refined elegance, from his expensive shoes to his neatly trimmed silver hair. Not even a billfold interrupted the cut of his expertly tailored suit. The General no longer bothered to carry cash, relying on the aides who shadowed him to discreetly settle any bills. Thanks to the private barbershops he had installed in his home and office, his appearance was perpetually impeccable, his nails always manicured. His old pinkie ring had been replaced by a new ring embossed with a flaming sword—the emblem of General Eisenhower's D-Day team, the Supreme Headquarters of the Allied Expeditionary Force, and an ever-present reminder of his service. Since the war, all ninety thousand employees of RCA and NBC had been instructed to address him as General Sarnoff. At home, even his wife and three sons had taken to calling him "General."

Sarnoff had responded to Armstrong's federal lawsuit by recruiting a legal army of similar heft. Behind Cahill sat William Davis, a top patent expert from the boutique firm of Davis, Hoxie, Faithfull & Hapgood. Next to him was Sarnoff's own Delaware expert, C. S. Layton, of Richards, Layton & Finger, also a Wilmington law firm. In all, ten lawyers crowded around the table, waiting for the interrogation to begin.

With his dimpled chin, glasses, and round, cherubic face, McCormack didn't look terribly intimidating as he began the deposition. He used his milquetoast demeanor to his benefit, and planned to start off the day's questioning gently in order to construct a detailed record of David Sarnoff's actions. Once the General committed to a specific story under oath, then McCormack could plot his attack on his narrative, exposing inconsistencies and falsehoods.

"General Sarnoff, will you state your residence and occupation?"

"My residence is 44 East 71st Street, New York City. My occupation is that of chairman of the board of the Radio Corporation of America and of the National Broadcasting Company."

"I would like to put on a brief statement of your career. I wonder, perhaps, if you could make that statement yourself, commencing with your first job at the Pearl Street office of the Marconi Co., which, I think, was in 1906?"

With that, the General recapped the early years of his forty-eight-year career, McCormack working methodically to put together a timeline of Sarnoff's actions during his long history with the airwaves and Edwin Armstrong.

The General fired back answers with unusual directness, showing off an impressive memory for events long ago and a surprising willingness to share details. McCormack tried to keep his witness from digressing, a process Sarnoff complicated by repeatedly veering away from the bare facts of his business relationship with the inventor to discuss the personal connection he shared with his old friend across the table, the friend who now blamed him for ruining his life.

Sarnoff took the opportunity to lay out his personal feelings toward Major Armstrong after McCormack opened the door by asking Sarnoff to describe "the nature of your contact with the Major" in the early 1920s,

"They were very friendly and close. We saw each other very

frequently, either in my office or in my home. We were close friends. I hope we still are."

Armstrong, sitting across the table, said nothing.

WHILE MCCORMACK VIEWED most corporate clients' dispassionately, he had developed a personal connection with Major Armstrong and his cause. Reading through the record of Armstrong's fight over FM radio made it nearly impossible not to sympathize with the bitterness and injustice the inventor felt. Wary and experienced though Armstrong had been in the postwar debate over the fate of FM radio, in retrospect he had not been nearly suspicious enough.

As McCormack saw it, what had enraged his client the most was not the FCC's imperviousness to engineering logic—bad as that was—but its utter lack of accountability for its mistakes. Nothing infuriated Armstrong more the failure of Kenneth Norton or Paul Porter to pay any price for crippling the young industry. Chairman Porter left office after just thirteen months on the job, going on to a lucrative career as a Washington lawyer and a founding partner of Arnold & Porter. Norton continued to work for the government as a radio propagation expert.

In one of his calmer moments, Armstrong summarized the problem this way: "There was one thing left out of the law; that was a requirement that [the FCC] be guided by the weight of the evidence. They have proceeded on the basis of evidence in favor of what they wanted to do, regardless of whether it constituted the weight of the evidence, and there is no correcting check on that power. I think that there is the heart of the difficulty."

The FCC's mistreatment of FM and its lack of accountability had sent Edwin Armstrong spiraling back into the rage and depression that had marked his years of litigation with Lee de Forest. By that point, AM radio interests had taken over 593 of the 700 FM stations on the air and mostly used them to rebroadcast the AM stations' programs—giving listeners little incentive to buy an FM set. None of the independent FM stations turned a profit and many were nearing bankruptcy. Giving up on ever winning justice at the FCC, he turned to Congress for relief in 1948. At Major Armstrong's prompting, both the House and the Senate launched hearings investigating the FM industry and the FCC's mistreatment of it.

At the same time, Armstrong expanded the scope of his battle, turning his attention and scrutiny on David Sarnoff, whom he held responsible for all his troubles with the FCC. Armstrong also began to publicly accuse Sarnoff and RCA of wrongdoing. He told the Senate committee looking into FM of the day in 1933 when he picked up Sarnoff at Rockefeller Center and drove him up to Columbia to show him FM radio in action. He described Sarnoff's reaction to the demonstration, including his comment that "this is no ordinary invention, this is a revolution." Then Major Armstrong leveled a direct accusation: "From that day till this, you will find the hand of the management of RCA working in the background against the advancement of FM broadcasting."

Armstrong argued to Congress that Norton's testimony to the FCC must have been part of a conspiracy, pointing to the alteration of Norton's classified testimony to make it look as though he had provided a satisfactory defense of his sunspot nonsense. The commission's new chairman, Wayne Coy, had assured the Senate panel that he had investigated the allegation and that even though he hadn't been able to figure out exactly who at the FCC had altered the public summary of Norton's confidential testimony, there was no reason for concern: "I am certain that wherever the change came from, it was first approved by staff as a proper answer." Any mistakes his predecessors made, Coy continued, had been nothing more than well-intentioned accidents. "In my investigation of the matter I have found difference of opinion and judgment on the technical matters involved, but I have not found evidence of dishonesty."

Armstrong's case against Sarnoff also fizzled, too technical for the lawmakers to follow and too circumstantial to convince them of Sarnoff's guilt. The RCA executives that Sarnoff sent to testify also mounted a compelling defense, cataloging all the things Sarnoff and RCA had done to help the technology as well as many of the things CBS and the local AM radio stations had done to harm it. To win Armstrong's anti-trust case, McCormack would have to prove two things. First he needed to show that his client's technology had been sabotaged, which promised to be the easy part. Second, he had to prove that it was David Sarnoff who had orchestrated the sabotage, and not the owners of the local AM radio stations who had far more to lose from a vibrant FM radio industry.

Only after the congressional hearings petered out in the spring of 1948 had Edwin Armstrong walked into Alfred McCormack's office in search of another way to carry on his fight. The two men, with the help of Armstrong's other lawyers, drafted a federal lawsuit accusing the Radio Corporation of America of patent infringement and the "intent to monopolize" the radio industry, a lawsuit specifically designed to force David Sarnoff to personally confront allegations that he had played a central role in hobbling FM radio in order to protect his own empire.

SO FAR THE TRIAL had been a slog. Simply gaining access to Sarnoff's memos and letters relating to Armstrong's inventions took months. RCA lawyers kept Armstrong testifying under oath for seventeen days over the course of four months before McCormack asked the judge to limit the "very detailed and apparently interminable examination on irrelevant matters." (The judge agreed to limit the questioning of Armstrong to another eight days.) And endless parade of witnesses followed in the ensuing years, with the federal judge taking the unusual step of appointing a "special master" to oversee taking of depositions that had slowed to a snail's pace. Sarnoff's testimony today would bring the told number of pages in the deposition transcript to more than 12,000.

To Armstrong's great disappointment, his army of lawyers hadn't been able to turn up an ironclad piece of evidence proving Sarnoff's betrayal. Still, McCormack and his team assured him they didn't need a smoking gun to win, just enough circumstantial evidence of motive, opportunity, and actions (and in some cases, inaction) to convince the court of Sarnoff's guilt. "The course of conduct that is attributed to the defendants," McCormack had told the judge in court, "is first a failure to put these inventions into use and to give them to the public; second, direct opposition to them and an attempt to suppress them and finally, after its efforts in that respect [failed] an adoption of the invention as their own and inducing others to infringe the patent." With Sarnoff's turn on the stand, Armstrong was confident that his legal team would soon possess enough evidence to show that Sarnoff had sought to stifle FM radio and rig the rules in Washington.

After a brief break, McCormack restarted his interrogation. David Sarnoff insisted that he had never done anything to destroy his friendship with Armstrong and that he had not wanted it to end, a position wildly at odds with the story Major Armstrong told. So why would Sarnoff tie himself to a blanket denial rather than just waiting to defend against specific allegations, as John Cahill and the rest of RCA's legal team surely would have preferred? For now, McCormack could only file that question away and go back to slowly working to set his trap. After a long review of the two men's professional relationship from 1910 into the 1920s, Colonel McCormack began to explore the issues at the heart of his case.

McCormack: "Do you recall, after your return from Europe in September, 1934, [that] Major Armstrong came to see you about the FM matter?"

Sarnoff: "Again, I can't recall the date, but I do recall that Major Armstrong came to see me on quite a number of occasions about the FM matter."

McCormack took out a copy of testimony that Armstrong had given in a recent Washington hearing and began to read a section, quoting Armstrong:

"He asked me why I was pursuing this development so hard. I told him there was a depression on and that this was the sort of thing I thought the radio industry needed to put life into it. I will never forget his reply. He said: 'Yes, but this is not an ordinary invention, this is a revolution.' I said I thought that was all the more reason to push it, whereupon the subject was changed."

McCormack stopped reading the transcript and addressed Sarnoff directly: "Do you recall having such a conversation?"

"Yes, I do."

Reactions to Sarnoff's answer flickered across the faces of the nine lawyers seated around the table. Defendants almost always plead a spotty memory, especially after twenty years. Instead, Sarnoff seemed almost eager to confirm a critical piece of Armstrong's case. Sarnoff's lawyer also seemed concerned, rising from Sarnoff's side of the table to suggest that his client be given a chance to give a more complete answer.

McCormack, still the picture of politeness, deferred to his witness.

Sarnoff: "I recall having made that statement, and I would like to define what I meant by 'revolution,' if I may. In order to do what Major Armstrong wanted, it would have been necessary to have changed all the broadcast transmitters in the United States and all the broadcast receivers and to have provided methods of interconnection [the telephone lines that send radio programs from city to city] that could take full advantage of the higher fidelity, because the telephone wires were limited in the frequencies they would pass.

"To substitute FM for AM, which was then used in broadcasting, would have constituted a revolution, that is a commercial revolution, an operating revolution, and you couldn't do that overnight. It would have obsoleted all the apparatus that was in the hands of the public, all of the apparatus that was in the process of manufacture and all of the apparatus installed in the many broadcasting stations. That, of course, is no reason for not adopting a new system, if the new system is so vastly superior to the one that it replaces."

Sarnoff continued: "I have always believed in progress in the radio art, and I am not afraid of obsolescence, as I think the record shows, but when one engages in that character of commercial revolution, one must be certain that the difference in results justifies the revolution. I was not at a point then where I knew enough to say that the difference made by this wideband or FM or the system that Major Armstrong was promoting, was so great in degree as to justify obsoleting the entire system of broadcasting in the United States. That is what I meant by 'revolution.' "

This time, before either side's lawyers could jump in, Phillip McCook, who had served on the highest appeals court for New York state and was overseeing the deposition, grabbed the opportunity to suggest a break.

McCormack and his team quickly assessed the meaning of Sarnoff's speech, and the hints it gave about RCA's defense. General Sarnoff had done them the favor of tying himself to a specific state of mind, claiming that his attitude about FM had not been hostile but rather uncertain. While Sarnoff claimed to welcome obsolescence, the idea that he'd happily allow Armstrong to blow up RCA and NBC's AM radio business strained credulity. McCormack needed to demostrate that the open-minded attitude to which Sarnoff just testified was a lie.

After the last lawyer returned from the break, McCormack started up again. He walked Sarnoff slowly through the mid-1930s, focusing particular attention on the day in 1933 when Armstrong demonstrated his new invention, FM radio, for his friend. McCormack then retraced the subsequent few years, contrasting the way technical evidence of FM's superiority mounted even as RCA and NBC's interest in promoting the technology appeared to dissipate. The way McCormack was attempting to frame the narrative, Sarnoff's betrayal had been so subtle that it had taken the Major years to realize what was happening. The first major clue to RCA's plans, McCormack intimated, was the behavior of Ralph Beal, a top RCA engineer whom Sarnoff had instructed to help Armstrong "make FM go."

In the mid-1930s, as experiments with FM provided a series of startlingly impressive results, Beal slowly stopped coming around. McCormack zeroed in on that point: "Major Armstrong has testified, I believe, a number of times, that after Mr. Beal talked with him informally on February 5, 1936, Mr. Beal never returned nor did anyone else."

At this, the so-far-unflappable witness responded, with evident emotion.

"If he said Mr. Beal never returned, I am not going to challenge Major Armstrong's statement. However, I think the Major is mistaken if he says no one else, because I think he may recall that there were a series of efforts, stimulated by me, to have emissaries between Major Armstrong and myself. First to see what could be done to improve the unfortunate atmosphere which somehow grew up between Major Armstrong and the RCA, and his apparent impression that I was the man that was not doing the things that he thought ought to be done."

As he spoke, Sarnoff began to make his case directly to the old friend sitting across from him, legal testimony lapsing into a personal appeal to reexamine how their relationship really came apart. Catching himself, Sarnoff turned his focus back to McCormack.

"When you go to negotiate with a man, certainly, I needn't tell a gentleman of your experience, that the first thing you have got to do is set an atmosphere for negotiation. And the list of emissaries is legion, beginning with Mr. Gano Dunn, a director of my company and a great engineer and a gentleman in whom Major Armstrong had great confidence, and in whom I still have all the confidence I ever had, and a friend of Major Armstrong's . . ."

McCormack had had enough. Gano Dunn, former president of Cooper Union and winner of the Edison Medal, was nearly as accomplished an engineer as Major Armstrong. McCormack knew his client had a stubborn streak. He didn't need to hear Sarnoff testify about its details. "May I interrupt you just a minute? I am glad to have all this information. I want to get it, but for the moment, if you don't mind, I'd like to go on with Mr. Beal."

"If you don't mind," Sarnoff replied, bulling ahead, "I would like to complete my answer, because the implication of your question was that Major Armstrong–that we didn't seek to make any arrangement with Major Armstrong."

McCormack cut him off, his earlier civility gone: "I intend to conduct this in an orderly fashion," he said, speaking over Sarnoff and demanding that Judge McCook instruct the witness to answer the questions McCormack asked.

Back in control, McCormack moved away from the personal relationship and returned to the task of proving that Sarnoff and RCA had long known just how superior FM was to AM. Slowly circling back on his prey, he began contrasting glowing reports from RCA's own archives with the dismissive letters that passed between Beal and Sarnoff during the same period. In one letter, the RCA engineer told Sarnoff that he doubted FM could ever overcome the advantages of the current AM technology. Colonel McCormack read aloud from Beal's letter to Sarnoff: "'I personally regarded the commercial and economic disadvantages as so adverse that I was very doubtful whether his system could prevail over amplitude modulation.'"

"Did you," McCormack demanded, "ever do anything to disassociate yourself from that view, as expressed by Mr. Beal?"

Cahill jumped in to object, but McCormack demanded Sarnoff answer the question, arguing that it cut to the heart of Major Armstrong's case. Judge McCook sided with McCormack and instructed Sarnoff to answer.

"I don't know what I did," Sarnoff replied, now clearly annoyed. "You assume that the only purpose in life I had was to just deal with this FM thing. I was a busy man at the time. I left the negotiations and explorations to my subordinates. I am not in the habit of asking them to report on what he said and she said every time they have a conversation.

"I did not find any necessity for 'associating myself' or 'disassociating myself' with anything, because Major Armstrong knew me well. He knew he was always welcome. It was as much his obligation to call on me as anything else, and if he kept away, why, I kept away."

McCormack: "Do you recall that the last conversation you had with Major Armstrong prior to 1936 was a demonstration in the summer of 1935, in which you told him that after Dr. Baker [a top RCA engineer who had worked with Armstrong on FM] returned from South America, you would have him get together with Armstrong and make FM go?"

"I don't remember the last statement, but I do remember speaking with Major Armstrong just before I went to Europe. I think I did discuss Dr. Baker with him and telling him I would get Dr. Baker to head it up, and Dr. Baker did, but I was not, myself–and I am trying to emphasize this–following up the detailed steps of the situation.

"It was a situation where Armstrong had as much direct access to my engineers as I did. He was constantly in touch with them. He talked with them. He had free access to the organization, to our laboratories, to our engineers. This was not a normal situation where you are dealing at arm's length.

"Up to that time, and up to the time that any differences of opinion developed between Major Armstrong and RCA, I had no limitations on what I would say to the Major or what he would say to me. I regarded him, as I have said, as much a member of the Engineering and Research Department of RCA as I would Baker or Beal or anybody else."

McCormack, having gotten under Sarnoff's skin, appraised his agitated witness with a practiced eye and decided the moment had come to dive into the personal questions of friendship, loyalty, and betrayal at the heart of the case.

"When did the differences of opinion that you have referred to occur?"

"I don't know," replied General Sarnoff, summarizing his testimony in a single sentence: "That was in Major Armstrong's mind, not mine."

CHAPTER 12

ROGERS

Princeton, New Jersey—1953

LAWRENCE ROGERS JR. WATCHED HIS FELLOW TV STATION MANagers stagger out of their shuttle buses. Several of them were feeling the effects of the previous night's festivities at the Waldorf Astoria bar, their hangovers now pounding in their heads with extra force following the two-hour bus ride to the middle of New Jersey.

Like the rest of the group, Rogers managed a local TV station that operated as an NBC affiliate. He looked forward to this trip every year, when NBC invited the station managers to New York for three days of high living at the Waldorf. In previous years, their second day had begun with a more hangover-friendly schedule: a short stroll from the Waldorf to NBC headquarters at Rockefeller Center for a 10 A.M. presentation by the network brass. This year promised something different. Just before leaving for New York, Rogers had received a letter from the NBC affiliates' volunteer leader, breaking the good news: "The General himself will meet with us."

While Rockefeller Center and Radio City served as the public face of RCA's power, the heart of David Sarnoff's empire could be found here, in rural New Jersey. RCA Laboratories had been founded in the run-up to World War II, and ever since then, the sprawling complex of low-slung buildings had been expanding into the adjacent farmland. Military contracts to build or research all manner of electronics—solar panels, radar jammers, transistors, and much more—fueled the rapid growth. The David Sarnoff Laboratories (as the complex had been renamed in 1951) was in essence a smaller version of AT&T's famed Bell Laboratories, where RCA's larger rival used a slice of its monopoly profits to fund all manner of pure and applied research.

Bud Rogers knew the area well from his days as a student at nearby Princeton University, an Ivy League pedigree that made him stand out from most of his seventy colleagues milling about the Sarnoff Labs parking lot. The other TV-station managers represented a cross section of American good ol' boys. Station managers such as Johnny Outler of Atlanta's WSB-TV and Howard Pill of Montgomery, Alabama's WSFA represented the classic southern variety. Ed Yocum of Montana's KGHL and Ewing Kelly of Sacramento's KCRA were similarly gregarious backslappers, minus the drawl. Not everyone fit the mold. Joseph Pulitzer, the publisher of the *St. Louis Post-Dispatch,* had come to represent the paper's television station. He stood in the parking lot, looking somewhat out of place.

Bud Rogers enjoyed hopping between these different worlds. After graduating from Princeton, he served as an artillery officer during the war, then moved to Huntington, West Virginia, to take a job at his new father-in-law's local radio station. (Like Joseph Pulitzer and many other longtime newspaper publishers, Colonel Joseph Harvey Long had wound up in the radio business shortly after the Federal Radio Commission started giving away licenses.) Introducing himself to the radio station's advertisers, who were spread out across central Appalachia, Rogers could be Bud from Huntington, West Virginia, a fella with a hillbilly name from a hillbilly town in a hillbilly state. Checking in to the Waldorf, he morphed into Lawrence H. Rogers Jr., scion of a well-to-do family, former army officer and Princeton man.

Rogers and his fellow station managers were well aware of the debt their booming new industry owed to David Sarnoff. Over the past quarter century, Sarnoff had poured $50 million of RCA's money into television research, doggedly ignoring others' doubts about TV's commercial potential. That investment was now paying off. RCA built more TV sets than any other manufacturer and the special wires Sarnoff had insisted on installing in NBC's studios two decades earlier during the depths of the Depression now served their preordained purpose, moving television shows around 30 Rock. These days nobody mocked David Sarnoff for being a "televisionary," that insult having matured into a compliment.

Rogers's main complaint about General Sarnoff—a gripe shared by his fellow affiliates—was the General's habit of focusing on technology at the expense of entertainment. Sarnoff's loyalty lay with RCA, the laboratories that created new technology and the manufacturing plants that built it, not with NBC, which produced TV programs he had no interest in watching.

He clearly appreciated the power that entertainment had over audiences, but General Sarnoff could never stomach the idea of a television star like Jackie Gleason earning more than a star research engineer working here in New Jersey, a comparison he often made whenever an NBC executive appealed to him for higher salaries to attract bigger stars.

The issue of Sarnoff's disregard for popular culture had come to a head recently at Rogers's West Virginia TV station, WSAZ. On Monday nights, NBC aired a long-running program devoted to classical music and opera while CBS broadcast a new situation comedy. To Rogers's mind, the contrast between *"The Voice of Firestone"* on NBC and *"I Love Lucy"* on CBS summed up Sarnoff's weakness as an entertainment executive. Lucille Ball's antics now attracted ten times the number of viewers as the NBC show—and still the General refused to cancel his high-minded but low-rated competitor. He also ordered NBC not to copy the CBS strategy for attracting TV's new stars. (To win the rights to *I Love Lucy,* longtime Sarnoff rival Bill Paley had sold a majority ownership stake in the show to a company the comedienne owned, greatly increasing her effective pay rate by letting her avoid postwar personal-income-tax rates that topped out near 80 percent.) To General Sarnoff , it made more sense to go hire a different actor, since to him comedians were interchangeable.

At today's meeting at the David Sarnoff Laboratories, the NBC affiliates would get the opportunity to air their complaints about NBC's programming. An article in the current issue of *Broadcasting* noted that the station managers were pleased that for the first time in recent years they would dealing directly with the top decision maker, "one who can give on-the-spot answers which, it has been their complaint in the past . . . had to be qualified and delayed while network officials consulted with top management of the parent RCA."

RCA aides ushered Rogers and his fellow affiliates inside, then directed them to a small auditorium. An array of the very latest in television technology, including several large color sets, sat onstage. Bud Rogers took a seat in the third row, watching as David Sarnoff walked onto the stage. The General was grayer now, at sixty-two, and missing his usual cigar, Rogers noticed. Otherwise, David Sarnoff appeared to be his old self.

The General exuded an aura of power, which people tended to notice even more than his elegance and wealth. "All 5 feet 5 inches of his bullnecked, bull-chested figure bristles with authority and assurance," *Time* magazine observed in its recent cover story on Sarnoff. "His chill blue eyes shine with impatient energy, his boyish, scrubbed-pink face radiates cockiness."

And yet, for all the evidence the morning's events seemed to offer of David Sarnoff's power over the new television industry, Bud Rogers knew better. His five years in the business had taught him that many things in the television industry were not what they appeared to be.

IT WAS HARD to believe that it had been only five years since Bud Rogers saw his first television set. Drinking scotch one night in a hotel bar, the radio ad salesman had been amazed to see Milton Berle's pie-covered face materialize before him on a small, black-and-white screen. Though Rogers had been reading about sending moving pictures through the air for years, he still couldn't quite believe what he was seeing. At that very moment, in NBC's studios at Rockefeller Center, five hundred miles away in New York City, pie was dripping off Berle's rubbery mug. Meanwhile, Rogers's fellow barflies were laughing along as if they were sitting in the studio audience.

It was a moment that Bud Rogers would point to for the rest of his life as the instant he decided to get into business with David Sarnoff: "I didn't see Milton Berle anymore. I didn't see anyone. I merely saw, in an instant flash of crystal clarity, where I would be going to spend the rest of my career—all my productive life. For the first time in my life, I was unqualifiedly certain at that moment that I knew what was going to happen."

Soon after his night in the bar, Rogers made his first trip to Rockefeller Center, asking rookie questions about how he might get into the television business. Only a handful of companies sold the equipment needed to operate a local television station—a transmitter, a large antenna, and TV cameras to film local news and events—with RCA dominating the field.

The real trick, Rogers learned, was not getting his hands on TV equipment but on a TV license, which would grant him the legal right to broadcast on a specified band of airwaves. Until he caught the TV bug, Rogers had paid little attention to the federal law governing the radio and television business (the Communications Act of 1934) nor to the agency it created to regulate the growing industry (the Federal Communications Commission). After speaking with a few Washington lawyers who specialized in representing clients before the FCC, he began to pay closer attention.

According to the law, the process should be simple and

straightforward. Ever since Herbert Hoover ushered the 1927 Radio Act into law, the basic principle supported by Hoover and Senator Clarence Dill had governed the nation's airwaves. (The Communications Act of 1934 replaced the 1927 radio law but retained its core provisions.) In theory, any citizen could apply to get a radio-station or TV-station license at no cost, just by asking. Above all, the law made clear, a license to broadcast electromagnetic waves of a particular length, which in essence is all a radio- or TV-station license is, was not to be considered private property. Instead, the government would lend out the licenses at no charge, and if multiple applicants wanted the same license, the FCC would judge who should get the first turn by assessing which licensee would better serve the public's needs.

Learning about license law gave Rogers his first lesson in the difference between the way people pretended the television industry worked and the way it actually did. Secretary Hoover, Senator Dill, and the other drafters of the law had failed to anticipate how their system of kindergarten-style turn taking would work in the real world. The cost of building a radio or TV station was nearly impossible to justify if the station could be run for only a single year, an economic reality overlooked by the law. As a result, the Communications Commission invariably ruled in the years that followed that the licensee currently running the station was also the best positioned to serve the public interest during the next license period. "Renewal expectancy," as it came to be known, effectively gutted what Hoover and Dill had been certain would be a key protection against license holder's gaining too much power.

Meanwhile, the FCC continued to pay lip service to the notion that station owners did not own the airwaves, even as it worked to slowly transform those licenses into what increasingly looked like legal title to a property—the right to exclusive use of those airwaves in perpetuity.

What most interested Bud Rogers was a simple question: Who ended up winning the original license grant? The law instructed the FCC to evaluate applicants based on "the public interest, convenience or necessity," a broad mandate that effectively gave the FCC the ability to pick whomever it liked for whatever reason it chose. The commission's one consistent criterion for license applicants, Rogers's lawyer told him, was proof of adequate financial resources to actually construct a television station. Beyond that, the criteria it used to select winners were opaque. On several occasions, the commission cited one reason for handing out a license (such as the winning applicant's extensive ties to the local community), then used the exact same reason to justify making the opposite decision a few months later.

Rogers also learned the story of how the FCC had designed the American television dial, an effort that had been part of the same postwar airwave allocation process that crippled the FM industry. Chairman Porter and his allies originally designed a TV dial with just thirteen channels. (The FCC decided to take away channel 1 shortly thereafter and reassign those airwaves to users of mobile radios, leaving TV dials in the United States numbered from channel 2 to channel 13.) Still, Porter promised, even twelve channels would be enough to create a total of 405 local TV stations across the country, and up to seven channels in big cities like New York, Chicago, and Los Angeles.

Critics, Edwin Armstrong among them, scoffed at the FCC's TV plan from the start. In a country as large as the United

States, 405 stations would not be enough to provide competition in many cities or to reach rural areas of the country. Worse still, Armstrong and his allies warned, the interference problems that the FCC had been so obsessed with in the FM radio fight actually did exist in its new TV band and would make it impossible to create even 405 stations. Sure enough, neighboring television channels soon began ruining each other's pictures. Signals from channel 4 in Detroit flew across Lake Erie, causing black slats to appear across the images being broadcast by channel 5 in Cleveland. By the time Bud Rogers got interested in trying to win a television license in 1948, the FCC had already run out of new licenses that wouldn't interfere with the 100 stations it had already approved— a quarter of its promised allotment of licenses.

As Rogers began to figure out the peculiarities of the FCC's plan, he realized that he was in luck. His home in Huntington, West Virginia, was in one of the few regions in the country where the television airwaves had not been given away already. Skeptics cautioned the TV-obsessed salesman to temper his enthusiasm, pointing out several obvious reasons nobody had bothered to apply for a TV license already. Huntington's population totaled all of eighty thousand people, and the surrounding region in Central Appalachia was infamously impoverished. Many people didn't even have electricity in their homes, let alone the means to afford a television. Furthermore, Rogers learned, television signals travel poorly in mountainous country, making building a TV station in the Mountain State, as West Virginia was nicknamed, even more ill-advised.

Rogers plowed ahead anyway, confident that he knew several things about his home that the TV folks in New York didn't.

Huntington, West Virginia, sits on the far western end of the state, smack in the middle of the giant Allegheny Plateau. That part of the Mountain State is not mountainous at all, and the giant plateau also extended across much of southern Ohio and eastern Kentucky. Studying topographic maps convinced Rogers that his idea could work. A TV signal broadcast from Huntington would reach far beyond the city limits to cover large parts of all three states, providing a far bigger market for viewers and for the sponsors who would want to reach them. Rogers figured this remarkable opportunity existed precisely because it seemed like such a dumb idea at first. With the help of his Washington lawyer, he applied for a television license and soon found himself in possession of a construction permit to build West Virginia's only TV station, WSAZ.

Even with that victory, many self-proclaimed experts continued to warn Rogers against wasting money on the station. On one visit to New York to shop for equipment, he met with one of television's best-known technical minds, who laid out the reasons why the new venture was doomed to fail.

Allen DuMont, a television pioneer who had made his fortune designing high-quality TV sets in the 1930s, told Rogers that the engineering incompetence that underpinned the FCC's channel allocation scheme was so appalling that, in order for the infant television industry to thrive, the current system would have to be junked. Twelve VHF channels could never offer more than one or two TV stations in most U.S. cities, he said, and no coverage at all for much of rural America.

The limited number of television channels had hit DuMont's television network especially hard, effectively leaving room for NBC and CBS to become the country's two television networks. Unless a city had more than three stations, no local broadcaster would choose to affiliate with DuMont's network, which could only afford to produce only low-budget dreck like *"Captain Video and his Video Rangers."* (That show revolved around characters in outer space who spent much of their time watching old westerns, an awkward plot device meant to keep production costs down.) The other second-tier TV network, the American Broadcasting Company, found itself in similar straits.

DuMont, the sort of play-it-safe type who wore both a belt and suspenders ("a habit that is popularly, and in his case rightly, regarded as an infallible index to character," noted a *New Yorker* profile of the engineer), sat Bud Rogers down and delivered a stern warning. Competing technologies offered obviously superior alternatives to the current setup. One technology, backed by Westinghouse, could deliver a dozen channels to virtually every American home by using decommissioned bombers from the war to create a sort of 1940s-version of satellite TV. Signals were sent up to the plane, which then broadcast them across a far greater area than any land-bound antenna could reach. During the 1948 Republican National Convention, a single plane used the "Stratovision" technology to broadcast the event across nine states.

DuMont told Rogers that even if the FCC ended up rejecting Stratovision, it would still be forced to relocate the TV dial just as it had relocated the FM radio dial. The Communications Act charged the FCC with maximizing the use of the airwaves, so sooner or later the commission would have to shift the TV band, allocating a larger slice of airwaves that would allow for more competition and, in the process, making early

TVs and TV stations obsolete. Did Rogers really want to follow in the footsteps of the FM pioneers who built stations designed to use wavelengths the FCC took back overnight?

SHORTLY AFTER ROGERS DECIDED to ignore DuMont's advice to give up on the TV business, the first part of the engineer's prediction came true. The FCC officially rescinded Chairman Porter's 1945 plan for TV, admitting that it would never provide adequate numbers of channels in American homes. The original decision to shoehorn TV into twelve interference-prone channels had relied on faulty engineering, just as Dumont, Armstrong, and countless other engineers had warned the agency. The FCC announced that it would stop authorizing any new TV channels and work on a new plan that would solve its old plan's mistakes.

In Huntington, West Virginia, Rogers could hardly believe his luck. For all that Allen DuMont understood about television technology, it was clear the engineer understood nothing about the television business. "Exactly 108 TV stations or authorizations were in existence, and they were all faced with the prospect of government protection for an indefinite number of years of monopoly—or near-monopoly—operation. For anyone with eyes in his head to see with and half a brain to think with, this would be the equivalent of turning the children loose in a candy factory!"

Rogers's station, WSAZ, hit the air on November 15, 1949, the sixty-fourth station to begin broadcasting in the United States. Like most local TV stations, it filled its time slots in two ways, with its own live programming (such as local news and sports) and by affiliating with a national network in order to broadcast it. For a station with a monopoly in its region, as Rogers's had, the choice came down to the Big Two, NBC and CBS, since the two smaller networks, DuMont and ABC, couldn't afford to produce a similar slate of high-quality shows. For Rogers, the decision was easy. CBS's head of station relations treated Rogers with barely concealed disdain, clearly unconvinced that the hick from West Virginia could ever build a worthwhile TV station. At Rockefeller Center, on the other hand, after giving Rogers a private tour of Radio City, an NBC executive presented him with a ready-made affiliation contract that Rogers' immediately signed.

The affiliation deal was a standard network contract, but Rogers could hardly believe the generosity of its terms. The contract reflected the peculiar balance of power that had transferred over to TV from the radio industry. While NBC and CBS called themselves "networks," the name didn't actually fit. Getting a TV show such as *"The Milton Berle Show"* delivered to a TV set in rural West Virginia took the cooperation of three rival interests. First, the "network" created the shows in New York. Next, it handed them off to AT&T, which owned a nationwide cable network capable of moving images and sounds between cities. Once the program arrived at the local television station, that affiliate would broadcast it over the air to nearby homes. The setup recreated the exact structure of the AM radio industry, where Sarnoff had been fighting with AT&T and the local station owners over how to divvy up the market ever since Rogers was a toddler.

The local television stations had been eager to mimic the radio industry's economic structure, for obvious reasons. In

1939, the FCC had shifted the industry's balance of power in favor of the local stations, outlawing most of the standard network-affiliate contract provisions that worked in the radio networks' favor. NBC fought the rules all the way to the Supreme Court, arguing the FCC's rationale—a desire to counteract the alleged monopoly power of the networks—was inappropriate for an agency that Congress never intended to make an adjunct enforcer of the nation's antitrust laws. In 1943, the Supreme Court sided with the commission, Justice Felix Frankfurter ruling that Congress had given the FCC wide discretion to protect the public interest, so if the commission felt it wise to issue rules aimed at restricting monopoly power, it could do so.

To traditional anti-trust experts, the FCC's rules were puzzling. In 1943, the year the Supreme Court allowed the rules to go into effect, the networks earned a pretax profit of $19 million compared to $47 million for the local stations—a situation that didn't seem to cry out for government intervention on the stations' behalf. Sure enough, the new pro-affiliate rules soon made it impossible for the radio networks to make any profit at all. A decade later, despite the introduction of TV, the local AM radio stations, continued to make solid profits, earning $45 million on $377 million in revenue in 1953. (The FM radio industry, meanwhile, had been effectively crushed by the FCC's regulations, taking in a paltry $2.3 million in revenue in 1953 and racking up losses so large that the FCC declined to release them.)

In the new television industry, the FCC's pro-affiliate rules combined with the scarcity of local stations to make the power imbalance between the networks and their affiliates even more extreme. Local television stations' profits shot up from nothing in 1950 to $50 million in 1953—and were rising fast. According to the FCC, the networks had gone from a $10 million loss to an $18 million profit in the same period, though Rogers knew that commonly cited number was extremely misleading. In fact, all the TV networks lost money, an embarrassing fact they covered up with the gusher of profits from their share in the local broadcasters' cartel. FCC rules allowed each network to own up to five local stations, and unsurprisingly the networks had picked the most profitable affiliates in the country's largest cities: WNBC in New York, KCBS in Los Angeles, and WABC in New York.

Most people never saw past the cultural power and giant revenues of the national networks, but Bud Rogers understood where the real economic power lay. "The forty-odd single-station market owners were in a position to demand literally anything they chose in the way of tribute from the networks and the advertising agencies who were trying to get their programs or their products exposed on TV," he wrote in his memoir. It helped that the FCC's "temporary" freeze dragged on for four years, from 1948 to 1952, allowing Bud Rogers, Johnny Outler, and the other members of the lucky 108 club to consolidate their hold on the markets they monopolized.

IN 1952, when the FCC finally announced plans to unveil its long-simmering plan to inject competition into the television industry, Rogers and his fellow broadcasters worried that their years of fat monopoly profits had come to an end. And indeed, to judge from the FCC's press release announcing its decision, the commission's plan would soon do just that. Since the regulators could find room for only a few more stations in the existing "Very High Frequency" band (which used airwaves over 1.4 meters long) it announced a plan to open up a vast number of channels in the shorter waves of the "Ultra High Frequency" band (which consisted of waves as short as 38 centimeters). Beginning in 1953, American TVs would go from having a dial limited to twelve VHF channels to a dial that also included another sixty-eight UHF channels. It had taken a while to work out a long-term fix to the market, the FCC commissioners said in announcing their UHF plan, but the new stations would finally unleash vigorous competition and solve the industry's problems once and for all.

The decision left VHF station owners ecstatic. Having spent the past four years getting to know how the TV airwaves worked in the real world, Rogers immediately understood that the picture being painted by the Communications Commission was a mirage.

The vital importance of a station's wavelength could be seen in the furious response of Bud Rogers to the FCC's plan to shift WSAZ from channel 5 to channel 8. (Though the FCC's 1952 plan focused on new UHF stations, it also created room for a handful of new VHF stations by moving around existing license holders.) The lower the channel number, the longer the airwaves it was allowed to use for its television broadcast, and longer waves were vastly superior at carrying TV signals over long distances. Channel 5, WSAZ's original assignment, used waves just shy of 4 meters long. Channel 8 used waves just 1.5 meters long. To maintain his existing coverage area, Rogers calculated, he would have to more than triple his power. Even if the FCC let him boost his power that much, he would still wind up with higher electric bills and lower picture quality across much of his territory. Even worse, he realized, the new station the FCC had slated for the state capital on channel 3 would have vastly better reach in his territory, thanks to its 5-meter waves: "Any nitwit could see that Charleston's channel 3 would dominate the whole area."

Rogers took his Washington attorney, Bernie Koteen, to see the FCC's general counsel. Like many communications lawyers in D.C., Koteen was a former FCC staff member himself, and the current general counsel, Henry Geller, was one of his former employees. Rogers, desperate to avoid being moved from channel 5 to 8, delicately dropped the phrase "*injunctive relief*" early in the conversation, warning the FCC lawyer that the switch would not take place without a long legal war. It would be easier to move WSAZ to channel 3, he suggested, and give the shorter waves to the new station in Charleston. By the end of the meeting the FCC lawyer had signed a letter allowing Rogers to do just that. In short order WSAZ vastly expanded its coverage area, including the entire city of Charleston.

Rogers's bitter fight to avoid being moved from channel 5 to channel 8 captured the absurdity of the FCC's plan to create viable competitors using the much shorter UHF waves assigned to channels 14 and up. Given the FCC's new rule, UHF stations would be lucky to reach enough viewers to stay in business, let alone make a profit. To Bud Rogers and his fellow station owners, this part of the regulators' plan was perfect. The local station owners had spent years worrying about the coming onslaught of new competitors that would follow the end of the long FCC license freeze. Instead, they now faced the only thing monopolists like more than a lack of competition—the illusion of competition.

As a final favor to the VHF stations, the commission's UHF decision also quietly authorized them to vastly increase their coverage areas by boosting their maximum allowable power from five to fifty kilowatts. Regulators also tripled the maximum height allowed for a broadcast antenna, to one thousand feet, magnifying the effect. Those rules promised to let the lucky 108 stations greatly expand the reach of their signals and widen the already yawning gap between themselves and any newcomers.

DAVID SARNOFF OPENED his speech to Bud Rogers and the other NBC affiliates by recounting his early years in the wireless industry, starting with his days at the Marconi Wireless Telegraph Company of America. The General had received a list of pointed questions from the affiliates ahead of time, covering their complaints over NBC's paltry daytime programming and a laundry list of other gripes. His talk touched on those challenges but focused on the need to prepare for the future that his researchers and engineers were now creating. The General followed no script, and for several hours Rogers never saw him refer to any outline or notes. Yet to Rogers's amazement the General never repeated himself, lost his place, or used an ungrammatical phrase.

"Of all the virtuoso performances I have ever witnessed in my life, this was the most outstanding," Rogers said after the speech. "For pure oratorical dynamism it has never been surpassed."

"He went through the vivid history of how all of us in that room had gotten there—mostly through following up on his own innovations over the past thirty years. And then he got to what was coming: Color TV.

"The General painted a word picture of the conversion of the entire TV spectrum to full color, and within a reasonably short span of years. To that end he pledged that, starting immediately, the NBC network would begin sending out some of its programs in "compatible color," so as to not interrupt or interfere with the reception of any existing regular programs. There would be massive promotion, and many of the new programs had been designed with color in mind. And, finally, there would be a crash program to get color TV receivers to the marketplace in time for the next Christmas selling season.

"Everyone in the room was going to be given the opportunity to sign up as a member of the world's first color-TV network before the day was over.

"There was much more to this presentation. And, of course, the audience lapped it all up like kittens and warm milk. When, finally, this marathon ad lib sales pitch came to an end, it was to a standing ovation including whistles and cheers that lasted for a full five minutes."

As the applause died away, Rogers heard the familiar Georgia drawl of Johnny Outler, who ran WSB in Atlanta: "'Call in the dawgs! Piss on the fire! The huntin's all over!'"

With that, Outler, Rogers, and the rest of the local broadcasters headed back to the buses for another night of partying at the Waldorf Astoria.

TWO YEARS AFTER Sarnoff's color-television presentation, as the glaring lack of competition in the television industry sparked calls in Washington for the FCC to correct its mistakes and to find an effective way to allow new stations on the air, Rogers and his fellow affiliates convened a meeting in Chicago to discuss a response. The group decided to create a new lobbying organization dedicated to stamping out the threat of new local stations. When a station owner asked what they should name their new lobbying organization, Nate Lord, of Louisville, Kentucky's WAVE-TV, quipped: "Why don't we just call it what we are: The National Association of Scared Fat Cats?"

After the station owners' laughter died down, they settled on a more artful name for a lobbying outfit designed to minimize competition: the Association for Maximum Service Telecasters.

CHAPTER 13

SARNOFF

New York City—1954

DAVID SARNOFF WALKED THROUGH THE LUNCHTIME CROWDS enjoying the unseasonably warm winter afternoon. Five blocks up Fifth Avenue from RCA's Rockefeller Center offices, he and the two aides who trailed him arrived at their destination. Above them, the Fifth Avenue Presbyterian Church's Gothic spire rose into the cloudy February sky. General Sarnoff turned up the red marble steps and led the way in to Edwin Armstrong's funeral.

Two nights ago, the sixty-three-year-old inventor had shocked everyone who knew him by leaping from his thirteenth-floor apartment in River House. The following morning a maintenance worker spotted his shattered body, dressed in a warm winter coat, scarf, and gloves at the base of the building. Today's newpapers carried stories blaring the news of the famous inventor's mysterious suicide. His funeral was set to begin at 1 P.M.

The news that his old friend had jumped from his apartment had stunned Sarnoff, not least because he had never known the Major to quit a fight. Many of his fellow mourners, Sarnoff knew, had remained close to the inventor and were undoubtedly familiar with the accusations that Armstrong had leveled at him in recent years. Many, no doubt, believed that Sarnoff himself had driven Major Armstrong to take his own life.

Major Armstrong's mysterious suicide had thrown his long, tortured relationship with David Sarnoff into the public eye. In the previous decade, Armstrong's myriad and unrelenting attacks on Sarnoff, RCA, and NBC—including his lawsuit and the congressional investigations he had spurred—never managed to seriously damage Sarnoff's authority or reputation. Even Alfred McCormack and the sixteen other lawyers Armstrong had hired to prosecute his lawsuit, now about to enter its sixth year in federal court, had yet to prove David Sarnoff guilty of anything other than being a hard-nosed businessman.

Inside, the cavernous church felt nearly empty. Roughly 150 mourners clustered near the front of a space built to seat a dozen times as many. Across the ocean of empty pews, Sarnoff could see his former secretary, the newly widowed Marion Armstrong. She sat with the rest of Armstrong's family in the first few rows, below an unusual pulpit that jutted out over them like a ship's prow. Trailed by his two aides, Sarnoff walked down one of the two aisles into the Presbyterian Sanctuary, as the church's main hall was known. Not wanting to call attention to himself and risk an ugly scene, he slid into an empty pew, out of sight of the family.

Sarnoff's attention moved to the unusual pulpit as the preacher from Armstrong's childhood home of Yonkers, New

York, Thornton Penfield, walked up the carpeted steps and began to lead the assembled in prayer. The minister started off in standard fashion, reciting the Twenty-Third Psalm and the Lord's Prayer, and then began a pointed eulogy.

"To these strengthening and undergirding words from the Bible, I add some words which seem particularly appropriate for our thought here," he continued, his voice booming across the empty pews. "We thank Thee for men of great integrity, men who are unpurchaseable," the preacher declared. "His mind was brilliant and creative, but he kept his genuineness, his integrity of spirit. He was not for sale."

As Penfield continued, there was nothing for David Sarnoff to do but listen in awkward silence. He never called out David Sarnoff by name, but otherwise making little effort to disguise the identity of the man he believed had tried to buy off the noble inventor.

THE YOUNG IMMIGRANT with the endless store of self-confidence had grown into a corporate mogul known for taking a keen, almost obsessive interest in his public image. Several RCA executives in the press department now spent their days chronicling his accomplishments, which already filled over a dozen volumes of RCA history. Other staff members spent their time hunting down honorary degrees or other awards for which Sarnoff might qualify. Despite the General's fame and power, he seemed to covet even the most trivial honors. Recently, after reading about an award a friend had won from a local advertising council, Sarnoff forwarded it to his staff with a note: "Is that something for me? Whom can we explore it with?"

Larger honors became the subjects of elaborate lobbying campaigns. Even the title of General turned out not to be enough. Lately Sarnoff had become fixated on persuading the Pentagon to award him a second star and promote him from brigadier to major general. So far, this had proven an impossible sell. Even with his old boss, Dwight Eisenhower, now sitting in the White House, the military tradition against promoting officers not on active duty had thwarted his ambitions.

Catty members of New York's media elite mocked Sarnoff's constant quest for status. Others considered it part of an understandable need to prove he belonged—not an easy feat for a Jew in the WASPy world of corporate America or a Russian immigrant in Joe McCarthy's America. A top Sarnoff aide put it this way: "His appetite for praise was never sated, either verbally or in print. This was the assurance he seemed to need of the reality of his transformation from immigrant to industrial nonpareil." The General, observed David Lilienthal, a close friend and the former head of the U.S. Atomic Energy Commission, "doted on publicity as much as anyone I have ever known." Rivals such as *Time* magazine's Henry Luce made a sport of tweaking the General for his "vast pride."

Yet the portrait of a vain and power-obsessed Sarnoff missed the essential philosophy that had always driven him. For the past forty years, ever since that fateful night of January 31, 1914, when Edwin Armstrong demonstrated his new amplifier for Sarnoff at an antenna site in Belmar, New Jersey, Sarnoff had maintained his faith in the growing power of the airwaves and a demonstrated reverence for the inventors who could build new tools to control it. In their long fight, no matter how blistering were Armstrong's personal attacks against Sarnoff, the General never retaliated. Instead, he stuck to the position he had taken when Alfred McCormack asked him about it a year earlier: "We were close friends. I hope we still are."

In contrast to the Presbyterian preacher now thundering on, David Sarnoff traced Armstrong's troubles to a different source: his own obstinacy. A quarter century earlier, when Armstrong launched the fight against Lee de Forest that would become the longest patent lawsuit in U.S. history, Sarnoff had sent him a note composed around a lighthearted anecdote about a stubborn horse standing on the train tracks staring down an onrushing locomotive. Sarnoff took the side of the farmer in the parable, who observed of his doomed horse: "I admire his courage, but damned if I admire his judgment."

It was one of Sarnoff's many good-natured attempts to prod his friend into accepting a reasonable compromise. Again and again over the years, Sarnoff had urged the inventor to spend less of his time fighting legal and political wars over past inventions and more time discovering new ones. As his friend's habit of fighting idealistic and doomed battles continued, Sarnoff's gentle prodding had turned to straighter talk. In 1934, after Armstrong showed him his FM discovery, Sarnoff tried again in his letter encouraging his friend to focus on his future inventions instead of battling in court over proper credit for the past.

Sarnoff had always credited Armstrong with teaching him about the invisible waves, insights that had sparked his own success. "I probably learned more about the technical operation of receivers and radio from Armstrong than I did from anybody else," was the way he'd put it to McCormack under questioning. Undoubtedly, the biggest lesson he absorbed from

Armstrong was something closer to a philosophy, or even a religious faith: Always trust in the untapped potential of the airwaves and in the ingenuity of human beings to reveal new ways to use the waves in order to connect mankind.

As Sarnoff saw it, Armstrong's fatal mistake was abandoning that optimism, trading his focus on the future for a life of recriminations, victimhood, and isolation.

Armstrong's perception of David Sarnoff after his first FM demonstration—a businessman intent on protecting his empire by throttling the superior technology—was the opposite of Sarnoff's image of himself. Just two weeks after the fateful demonstration in 1934, he delivered a speech in which he celebrated obsolescence and pledged his devotion to the technologies of the future: "The total of the world's inventive enterprise and scientific resourcefulness has not eliminated one frontier of knowledge. It has merely pressed those frontiers forward a little more into the mists of the unknown, a territory still larger than the limited area of our knowledge," he said. "We have broken our earthly bonds and have started to hammer at a new frontier, vaster beyond all imagination than any within human experience."

No matter the existing industries a new invention might threaten, he told his audience, trying to delay the pace of progress was a mug's game. Tomorrow's wireless applications were certain to be more powerful and valuable than yesterday's. Only a pessimist or a fool would defend the past rather than rush headlong into the future, and David Sarnoff was neither: "Any attempt to freeze society and industry at a given time or point will be as ineffectual as undertaking to hold back the onward rush of time."

To David Sarnoff, it was one of the enduring mysteries of Edwin Armstrong's mind: Had he really believed that Sarnoff had been faking it all along? Sarnoff could understand why Armstrong believed that the Federal Communications Commission had rigged the game against him. A mountain of circumstantial evidence now supported that case. But Armstrong's antitrust case against RCA had stalled for the same reasons his congressional hearings had collapsed—because of all the evidence that directly contradicted the inventor's allegation that "you will find the hand of the management of RCA working in the background" to thwart FM radio.

In 1940, Sarnoff's company had backed Armstrong's application with the FCC to allow FM on the air. In 1945, the engineer in charge of RCA Laboratories had been one of Armstrong's staunchest allies in the fight to expose Kenneth Norton. When CBS backed Norton's plan to move FM radio to shorter waves, RCA and NBC publicly opposed it. The FCC's decision to weaken FM radio range through its "single-market plan" had been proposed by CBS and opposed by NBC. Yet none of that ever seemed to shake the inventor's conviction that the man who had crushed the FM radio industry had orchestrated the conspiracy from his office on the 53rd floor of Rockefeller Center.

Nor had Armstrong bothered to examine the market he claimed David Sarnoff had conspired to monopolize. True, the General had refused to make the declining fortunes of NBC's radio network the centerpiece of his defense, but anyone who read the trades, spoke to industry experts, or simply looked at the FCC's annual data could have understood how the AM radio industry had changed since the FCC's regulations designed to aid local radio stations went into effect a decade earlier. The NBC radio network was set to post another loss in 1954, yet Edwin Armstrong never stopped to question the wisdom of accusing a money-losing business of running an illegal monopoly.

FROM THE PROW-SHAPED PULPIT, Reverend Penfield began to wrap up his eulogy. He veered away from his thinly veiled indictment of Sarnoff to celebrate Armstrong's unflagging obstinacy—precisely the part of Armstrong's character that Sarnoff had always found so troubling. "We thank thee for characters of strength who dare to carry on when there is little hope of fulfillment," the reverend proclaimed, launching into a lofty simile that painted the inventor as a brave mountain climber who fights to conquer "unconquerable obstacles."

"The saints," the minister concluded, "are only ordinary people who carried on."

As the choir launched into a hymn and the mourners filed out of the church, Sarnoff and his RCA colleagues slipped out quietly, drawing looks from a few newspaper reporters but avoiding the notice of Armstrong's relatives.

It was a shame that Major Armstrong had given up the fight when he did, especially because RCA scientists had started work on a new invention that Sarnoff believed would finally make it possible to launch an effective attack on both the phone monopoly and the local broadcasters' cartel. If only Major Armstrong had heeded his advice and shown more care and patience in the fights he picked, perhaps the two of them could have teamed up to win one more grand victory.

For forty years David Sarnoff had remained true to the faith he and Edwin Armstrong had once shared, a belief in the unstoppable expansion of the airwaves' power to connect the planet. Yes, there were powerful forces dedicated to holding back the future, and yes, those forces would sometimes succeed—but only for a while. The key, as Sarnoff saw it, was keeping his focus on the future, compromising when necessary and attacking only when the moment was right.

As for blithely trying to "conquer the unconquerable"?

That was David Sarnoff's definition of stupidity, not sainthood.

CHAPTER 14

SARNOFF

East Windsor, New Jersey—1961

MANHATTAN'S SKYLINE SHRANK IN THE REARVIEW MIRROR AS David Sarnoff's limousine headed south on the New Jersey Turnpike. Forty-seven years had passed since he and Edwin Armstrong had traveled this very same route, heading to the Marconi Wireless Telegraph Company station in Belmar to test the young inventor's amplifier. Much had changed. A major turnpike now ran south from Manhattan. The jarring ride of his Model T–era car had been replaced with the air-conditioned luxury of his chauffeur-driven limousine. Sarnoff's round face, once deceptively youthful, now displayed all the wrinkles and spots of a man entering his eighth decade of life. His once-thick shock of brown hair had retreated to a semicircle of gray.

Still, the twenty-two-year-old chief inspector of the Marconi Wireless Telegraph Company was easy to spot beneath the exterior of the seventy-year-old corporate titan. David Sarnoff remained as captivated as ever by dreams of discovering new, cheaper ways for people to share information. In 1914, the twenty-two-year-old David Sarnoff focused on toppling the transatlantic cable cartel by using cheap waves in the air to leap over the cartel's expensive cables on the ocean floor. Now, a half century later, the seventy-year-old Sarnoff had become obsessed with a remarkably similar plan. The technology had taken longer than he had expected to leave the lab, but today the General hoped to confirm his suspicion that RCA's new weapon would prove powerful enough to defeat Major Armstrong's enemies—and his own. His limousine pulled off of the Jersey Turnpike and turned east.

Ever since his days traveling the eastern seaboard as a Marconi inspector, Sarnoff had delighted in opportunities to see the latest wireless inventions in action. These days that usually meant a limo ride for another eight miles to the east, to the David Sarnoff Laboratories in Princeton. Today, however, his limousine pulled into the parking lot of a new RCA facility. Above him, a large, windowless building rose into the air. The giant featureless box would have looked unusual almost anywhere, but here amid the New Jersey farmland it seemed especially alien. A giant sign running along the building's top edge offered the only clue to the work going on inside: "RCA SPACE CENTER."

LIKE MOST LARGE U.S. companies in the mid-twentieth century, RCA required its executives to retire at sixty-five. Five years earlier, when that deadline caught up with David Sarnoff, RCA's profits had been in a steep slide. A few disgruntled shareholders

thought this reason enough to ease the relic of the radio age into retirement. RCA's board declined, ignoring the policy and allowing its chairman to maintain his complete control of the management of RCA.

That faith had been richly rewarded. As the 1960s began, RCA was on a tear. Sarnoff's conglomerate now built a wide variety of cutting-edge, high-tech products and had vaulted it into the twenty-sixth spot in the Fortune 500, with revenues topping $1.7 billion. The company's new space division, RCA Astro Electronics, stood as an example of one of David Sarnoff's two major postwar triumphs. Before World War II, RCA split its manufacturing into sixteen major product groups, only one of which focused on selling gear to the government. This year, RCA would collect $600 million, just over one third of its total revenue, from government contracts. Since the end of the war, RCA had built the world's first weather satellite (for NASA), new sonar systems (for the navy), and a trio of sixteen-story-tall radar dishes designed to spot Russian intercontinental ballistic missiles at a distance of two thousand miles (for the air force).

Congress, spurred into action by the Cold War, the Korean War, and the space race with the Soviets, underwrote much of RCA's basic research. The U.S. Army funded an RCA study of materials that could convert sunlight into electricity, which led to the company's current effort to produce commercial solar panels. Government cash also helped pay for engineers designing new types of transistors, lasers, superconductors, and magnetic data storage. In all, almost twenty-two thousand RCA employees worked on government contracts, here at RCA Astro and at ten other facilities around the country.

The General and his stockholders were also enjoying the spoils of his long war to make RCA the dominant force in color television. In typical fashion, Sarnoff had declared the commercial viability of the new technology far too early. After convincing the FCC to accept RCA's compatible, all-electronic standard in 1953 and then winning commitments from Bud Rogers and his other affiliates to switch to color broadcasting, Sarnoff had expected the new technology to take off quickly. It hadn't.

After the FCC anointed RCA's technology as the standard, CBS's interest in upgrading its shows to color vanished. Cash-strapped ABC also refused to colorize. Why would it bother? Too few viewers in the 1950s had color sets. With RCA's color sets selling for $500, twice what a black-and-white model of a similar size cost, American consumers matched the other networks' lack of enthusiasm. To top it off, RCA's rivals in the TV-set-manufacturing business had no incentive to help solve the chicken-or-egg dilemma. Thanks to the $160 million Sarnoff had dumped into color-TV research, only RCA engineers knew how to mass-produce high-quality color picture tubes.

Sarnoff, confident as ever, charged ahead alone. He ordered RCA factories to begin converting to the production of color sets, betting that orders would materialize by the time they were built. He directed NBC to convert its lineup of black-and-white shows to color. By the late 1950s, reminders of Sarnoff's color crusade had become a part of the country's daily life. NBC switched its logo to a peacock with a multicolored tail. Then the network attached new "bumpers" to the beginning of its prime-time programs, short animations of a black-and-white peacock fanning its plumage to reveal a vibrant,

color rainbow. To make sure the vast majority of viewers still using monochrome sets didn't miss the point, a voice-of-God announcer enthused: "The following program is brought to you in living color on NBC!"

To Sarnoff, the all-out color push was a perfect example of why the world needed RCA. Only his company had combined the research talent, manufacturing capacity, network programming, and guts that it would take to turn the technology into a viable product for ordinary consumers. "In a big ship sailing in an uncharted sea, one fellow needs to be on the bridge," the General told one reporter, in between puffs on his ever-present cigar. "I happen to be that fellow."

It took until the 1960s for the technology to really catch on, as falling set prices and improving picture quality jump-started color TV sales. Now, in 1962, every major television manufacturer, afraid of getting left behind, built color sets with picture tubes purchased from their rival, RCA. ABC's chief, Leonard Goldenson, had caved as well, recently ordering up his network's first color show. (*"The Jetsons,"* a cartoon about a space-age family, marked the beginning of ABC's $50 million effort to convert its network to color.) Even CBS chief Bill Paley had abandoned his campaign of ridiculing the technology and had switched, regularly airing movies and other specials in color. In 1962, sales of color TVs were on track to triple to 450,000, with RCA factories producing 100 percent of the color picture tubes and 55 percent of completed sets.

RCA remained a long way from recouping the $160 million that Sarnoff put into developing color technology—a staggering investment that exceeded RCA's total earnings over the previous decade. Yet at the pace color TV profits were growing, that seemed like just a matter of time. Jack Gould, the *New York Times*' TV columnist, summed up Sarnoff's triumph this way: "Not only did Mr. Sarnoff overthrow the Columbia Broadcasting System's method of color transmission, he took on every last concern in TV broadcasting and set manufacturing to popularize color video in the home."

Perhaps the only one who was unsurprised by the phenomenal success of color TV was David Sarnoff himself. To him, it was just another confirmation of the lesson that he had learned from Guglielmo Marconi half a century earlier: when wagering on the future of a new wireless technology, always bet on the optimists—eventually they're going to be right. The only reckless gamble, David Sarnoff liked to say, was betting against the invisible waves. To a *Time* magazine reporter quizzing him about the future of communications, he put it this way: "He is a rash man who would limit an art as limitless as space itself."

IN KEEPING WITH another of his old habits, now that the general public was finally embracing color television, David Sarnoff had moved on, shifting his attention to a new invention that he believed would multiply the power of the invisible waves in the air once again. That quest was what had brought him here, to the converted warehouse in the middle of New Jersey.

These days, RCA's main businesses of manufacturing high-tech gear accounted for 76 percent of its revenue, split almost evenly between government contracts and sales to consumers and private businesses. The NBC network and its five hyperprofitable local stations brought in another 22 percent of revenue. (The company did not disclose the contribution each business line made to its total profits.) Today, however, David Sarnoff wasn't thinking about any of that. His attention was on the oldest and smallest of RCA's four main divisions, RCA Global Communications, which accounted for the final, and often overlooked, 2 percent slice of the RCA empire.

For the past forty years, RCA Global Communications had made a business of selling the original wireless app: transoceanic text messages. Within two years of RCA's founding, AM radio eclipsed this original line of business, which soon was all but forgotten. Still, it had continued to grow, year after year, decade after decade. Sarnoff's engineers regularly discovered new and better ways to bounce message-carrying waves off the ionosphere. After World War II, for instance, RCA Global Communications started selling "teletypewriters." Using the service, any company that owned a phone line and an RCA teletypewriter machine could send a text message overseas directly from its own office. First, the text traveled over the phone network to an RCA wireless station, where it was transformed into a burst of airwaves for the jump across the ocean. Next, the message was routed back onto the overseas phone network for delivery to a second teletypewriter. That device would spring to life and spit out the original message, only moments after it was sent. In essence, teletypewriters and "telexes" allowed businesses to run their own private telegraph office. These days, waves in the air carried twice as many messages to and from North America as all the wires under the sea.

In 1960, RCA Global Communications started offering a new communications service that eventually would become the dominant offering of the twenty-first century: a generic digital connection that customers could use for whatever purpose they wanted. Running at up to 1,200 bits per second, the newfangled service offered "the first customer-to-customer service capable of sending vast volumes of data overseas with great speed and accuracy."

For his next technology crusade, David Sarnoff intended to target a different network of earthbound wires: the telephone network owned by RCA's oldest and most powerful rival, AT&T. Until recently, even an optimist like Sarnoff would have branded such an idea impossible. Compared with the land-based phone network, old-fashioned radio telephone systems cost more to operate and could carry far fewer phone calls. Now, though, after forty years of bouncing messages off of the earth's atmosphere, RCA engineers were about to place a new wave reflector in the sky. Launching an orbiting spacecraft known as a communications satellite would allow—if optimistic projections proved true—transoceanic wireless links to move a million times as much information as RCA's current technology.

Satellite communications—the idea was strikingly similar to the one Sarnoff had proposed in his speech to Herbert Hoover's radio conference in 1924 and then fought for, with very limited success, for the next fifteen years. In the 1930s the FCC did authorize a handful of what it called "clear channel" AM stations, though it limited them to one-tenth of the power that had Sarnoff requested. True "superpower" AM radio stations could have reached huge sections of the country at all hours of the day, bypassing both AT&T's expensive phone lines, which delivered network-made shows to local stations, and the wildly lucrative local broadcast stations, which took the shows from AT&T and broadcast them to local homes.

Now satellites promised to make possible a sort of superpowered version of superpower radio, capable of delivering not just low-fidelity AM radio shows but TV programs and vast numbers of long-distance phone calls as well. Delivering NBC's TV and radio shows directly to affiliates would save Sarnoff's network the $20 million a year it paid AT&T to use its cables. Those savings would double RCA's profits.

Future generations of satellites—and Sarnoff knew this was the part Major Armstrong would have relished—promised to become powerful enough to beam TV and radio shows directly to people's homes, freezing out local affiliates altogether. Those local television stations now pocketed 88 percent of the TV industry's profits ($280 million for the local stations versus $31 million for the three national networks) and 107 percent of the radio industry's profit ($45 million versus a loss of $2.8 million for the networks). A study by Ronald Coase, a University of Chicago economist and future Nobel laureate, found that many VHF station owners were earning a mind-boggling two-hundred to three hundred percent return on their invested capital—*every year*. If satellites could circumvent the local station's bottleneck, it would flip those economics upside down.

INSIDE THE RCA ASTRO BUILDING, Sarnoff's top researchers showed off their progress on the new device he had come to see: RCA's first communications satellite. The partially assembled craft was a peculiar sight. Only thirty-three inches tall, "Relay-1" was shaped a bit like an oversized drinking glass. Its top end stretched twenty-nine inches across, while the bottom, designed to point earthward, was fitted with an unusual antenna. Small solar cells, responsible for keeping the satellite charged up in space, covered its exterior. By the time Relay-1 was completed, 8,215 tiny panels would produce forty-five watts of power.

A rotating cast of engineers showed off the latest equipment they were using to simulate the hazards RCA's satellite would face once it passed beyond the protection of the earth's atmosphere. The most spectacular piece of gear was a massive metal contraption that looked like a cross between a water tower and a bank vault. Bright yellow, inch-thick steel walls enclosed an empty cylinder twenty feet high and twenty-six feet in diameter, room enough for the fully built satellite. Once a nearby crane lowered the satellite into the container, a series of pumps removed nearly every molecule of air from the chamber, simulating the vacuum of outer space. Copper tubes around the outside of the chamber allowed the engineers to rapidly raise and lower the temperature inside the vessel, creating a cycle of searing heat and frigid cold that mimicked an orbit and that moved between the sunny and shady sides of the earth. This device had already helped to uncover a number of problems that would have turned the satellite into space junk. In one case, surprised RCA engineers found that once the pressure fell low enough, normally stable lubricants vaporized.

Nearby, David Sarnoff could see a smaller machine, this one closer to his own height, designed to subject space-bound components to brain-rattling vibrations. A high-power, oil-cooled "shaker" grabbed pieces of the satellite and jiggled them back and forth as many as ten thousand times per second. A separate "shock tester" simulated the effects of being blasted into space, subjecting the satellite's components to as much as 250 G's.

While Sarnoff was curious about the details of all the testing equipment, he was primarily concerned with how much information Relay-1, and its progeny, could move. For once, he recognized, he was not alone in his pursuit of an unproven but potentially revolutionary communications technology. The basic idea—and remarkable potential—of using a spaceship as an airwave reflector had been floating around ever since Arthur C. Clarke wrote an article describing how it could work in a 1945 issue of the magazine *Wireless World.* Clarke, a physics whiz who later built a career writing science fiction, proposed placing a satellite over the equator. A satellite placed 23,200 miles up would take exactly twenty-four hours to circle the planet, Clarke calculated, meaning that from the perspective of a person standing on the ground, it would appear motionless, always hovering overhead in the exact same spot.

From that distance, such a "geosynchronous" satellite would have a direct line of sight to one third of the earth, making it possible to bounce all sorts of information—phone calls, text messages, television shows—off the satellite and across the world's biggest oceans and continents. Clarke also predicted that launching a few satellites would be a lot cheaper than laying a web of undersea wires: "However great the initial expense, it would only be a fraction of that for the world networks replaced, and the running costs would be incomparably less."

Sarnoff had long been captivated by the potential of the tiny waves and followed research aimed at increasing their usefulness closely. Microwave technology (using waves shorter than one meter in length) could now beat AT&T's metal cables in both cost and capacity. Even the lumbering phone monopoly, which for eighty years had woven its network out of metal wires, now used microwave towers along some of its high-traffic routes, a fact Sarnoff was reminded of every time he drove down the New York Thruway and saw the telltale towers lining his path. Canada's national phone company had gone much further, building a coast-to-coast microwave system with enough capacity to ferry 2,400 phone conversations and two TV shows, such as *Hockey Night in Canada*, over a chain of 139 towers.

Using microwaves to communicate across oceans had been unthinkable until after the Second World War. Unlike the longer waves that Marconi glanced off the ionosphere, microwaves pass right through it, zooming off into outer space. Back in 1948, Sarnoff had written Secretary of Defense James Forrestal with an audacious proposal to use a line of microwave-equipped airplanes flying across the Atlantic as communications towers in the sky, each flying high enough to have a direct line of sight with its neighbors. "There should have to be 30 flights in each direction daily, and the schedules should be uniformly distributed to provide a continuous circuit," Sarnoff told the defense secretary. The circuit "could accommodate the following services: 1. Several hundred telephone calls. 2. An ultrafax system at 540,000 words per minute in each direction. 3. A television relay in each direction. 4. Several thousand teleprinter channels, each working at 60 words a minute in each direction." The Defense Department passed on the idea, but a dozen years later Sarnoff's interest in beaming microwaves across oceans burned brighter than ever.

At rival AT&T, interest in the potential of satellites was

similarly fervent. For the telephone company, the satellites offered a space-age solution to a jazz-age problem: how to get phone calls across an ocean. Despite all their practical triumphs and Nobel Prize–winning theoretical research, even Bell Labs' brainiacs struggled to send telephone calls through cables running beneath the ocean. In 1954, AT&T spent $35 million to lay its first transatlantic telephone cable, capable of carrying thirty-six low-fidelity phone conversations at a time.

John Pierce, the genius in charge of keeping all of Bell Labs' other geniuses in line, was an avid fan of bouncing microwaves off satellites. He assigned mathematicians to calculate the visibility of satellites in different orbits, ordered chemists to improve nickel-cadmium storage batteries for the satellites, and personally oversaw the effort to calculate how much power it would take to get a microwave to space and back. A single satellite, Pierce calculated, could carry a thousand phone calls and thus be worth as much as $1 billion. Even for AT&T, that was real money.

In its competition with RCA, the phone company had jumped out to an early lead. Just before dusk on October 28, 1959, Bell Labs launched a giant Mylar balloon coated in aluminum. When the balloon, one hundred feet in diameter, reached an altitude of 250 miles, it caught the rays of the setting sun and lit up like a mysterious orb, glowing like a huge fireball in the sky. Panicked calls flooded into NASA as citizens from Cape Cod to Charleston spied the bizarre sight. On that flight, Bell Labs' efforts to bounce a microwave off the giant balloon failed. On its next flight (this one announced to the public ahead of time), the company managed to hit the balloon with high-power microwaves, and then detected the waves' reflection in a laboratory in Massachusetts. Only one millionth of the beam's power reached the balloon, which then scattered the reflection in all directions. The waves that happened to reach a distant receiver were incredibly faint, roughly one quadrillionth the power of the original transmission, according to John Pierce's calculations.

RCA began working on the next step, building an "active satellite" that could act like a microwave relay tower. As Sarnoff explained the idea, Relay-1 and other "active satellites" would "act as a catcher-and-pitcher combination, catching an electrical impulse of signal, sending it through amplifying tubes to boost its strength, then pitching it on its way."

After an extensive competition, NASA selected RCA to build Relay-1. Sarnoff reveled in that victory: he had triumphed over his bitter rival, AT&T, and received more cash to fund RCA Astro's research into satellites. Sarnoff was anxious to extend that lead. Pierce, of Bell Labs, had said publicly that while satellites would be great for international communication, they would never be powerful enough to compete with his employer's network of metal wires and microwave towers on earth. Perhaps his skepticism was born of a lack of imagination, perhaps from an excess of company loyalty. Either way, for Sarnoff, it revealed a blind spot that made the idea of using satellites to attack the phone giant all the more appealing.

According to RCA Astro's engineers, there was every reason to expect that future generations of satellites would grow strong enough to shift the balance of power, and thus the balance of profits, between the national networks and the local stations. Sarnoff, hoping to act ahead of the shift, had discussed with his top aides the idea of unloading RCA's most profitable assets: the five local television stations NBC owned. The market value of each one of those stations now topped $30 million and, as Sarnoff pointed out, that was more than enough to launch a satellite into orbit. Building, owning, and operating the progeny of Relay-1 would transform NBC into the first broadcast network worthy of the name—a TV network actually able to carry its programs around the country.

SARNOFF'S LIMO HEADED back to Manhattan. Today, like so many times in his life, David Sarnoff had been lucky enough to get an early look at a discovery that would begin a new communications revolution. This time, however, Edwin Armstrong had not been standing at his side.

David Sarnoff was not the type to waste time on regrets, but it was impossible not to imagine what could have been in the two friends had remained allies. Launching an attack against AT&T's phone monopoly and the local broadcasters' cartel from space would be the final crusade of David Sarnoff's epic career, one that he would have to fight alone.

CHAPTER 15

JOHNSON

Washington, D.C.—1962

CROWDS OF SPECTATORS JAMMED THE PUBLIC GALLERIES OVER-looking the U.S. Senate, eager for the day's fireworks to begin. Normally the public paid no attention to debates over bills with names as dull as this one, and Vice President Lyndon Johnson would have preferred to slip the Communications Satellite Act of 1962 through Congress quietly. The proposed law restricted ownership of communications satellites to a new company that would be controlled by AT&T and was designed, its supporters said, to ensure that the new technology got off to a fast start. But below the dais where he sat, the vice president could see the group of dissident senators dedicated to making sure that didn't happen. Senator Ralph Yarborough of Texas, one of the group's leaders, began to hammer on Lyndon Johnson and the bill's other supporters, describing what he and his allies saw as a more nefarious motive for the proposed law: eliminating the mortal threat that satellites posed to America's communications cartels.

"It is rather depressing," said Senator Yarborough, "that, on the threshold of one of the greatest scientific discoveries in the history of the human race, we have a greedy little band that reaches out to Congress and says: 'Give it all to us. Let us have it.'"

Before leaving the Senate to become JFK's vice president, Lyndon Johnson had spent his happiest days as a politician in this room. As the Senate majority leader, he excelled at intimidation, becoming famous for his ability to bend other senators to his will. In his eighteen months as vice president, LBJ often grumbled to friends, his life in politcs had turned into a chore. Today offered a welcome change of pace. The United States Constitution gives a sitting vice president a single way to directly control the levers of power. In the V.P.'s secondary role as the president of the Senate, Johnson had the right to preside over the Senate's debates and, in the rare event of a tie, to cast a deciding vote. Today, for as long as Johnson sat atop the dais in the well of the Senate, the lawmakers addressed him as "Mr. President." LBJ, who had challenged JFK for the presidential nomination at the 1960 Democratic convention before agreeing to take the second spot on the ticket, preferred that salutation.

From the floor of the Senate, the drawl of Senator Yarborough continued, as he took his time describing the venality and hypocrisy he saw in the bill's supporters. Having given up hope of shaming the bill's supporters into switching sides, he and his allies now hoped to talk the bill to death.

According to century-old tradition, senators only filibustered a bill in exceptional circumstances. In recent years, a group of southern senators who bitterly opposed civil rights laws backed by President Kennedy had been the only ones to use the tactic. Defeating a filibuster was possible, Johnson knew, but it was a parliamentary maneuver so rare that he had never seen it succeed during his dozen years in the Senate.

That was the challenge that Lyndon Johnson had returned to the Senate chamber to take on: squash the filibuster and pass the bill. "Issues affecting the Senate's procedures and bound up with its hallowed rules always bring tension and high drama to the old chamber in the north wing of the Capitol," wrote one reporter watching the action. "And nothing so captures Washington's imagination as a filibuster."

Senator Yarborough's speech made it easy to see why. Before finally yielding the floor to a fellow filibusterer, he aimed an unusually pointed barb directly at the vice president (addressing him, according to tradition, in his current role as president of the Senate): "Mr. President, is this the council hall of the States, or has the Senate become the council hall of the corporations?"

SINCE THE DAWN of the satellite age, RCA's David Sarnoff had been the leading public enthusiast for the technology, often rhapsodizing about a communications industry remade by the new technology. Testifying before Congress, Sarnoff described how satellites could soon relay programs produced by NBC and its fellow networks directly to local TV and radio stations, Sarnoff pointed out, slashing the $50 million they now paid AT&T every year to use its wires for the same purpose. Soon after that, he predicted, satellites would become powerful enough to broadcast a vastly expanded selection of channels directly to people's homes, skipping the bottleneck now created by the shortage of terrestrial TV channels. He also predicted that satellites would eliminate the need for AT&T's monopoly on long-distance telephone calls. In the presatellite era, building multiple long-distance phone networks had been too expensive to seriously contemplate. That was no longer true, Sarnoff had testified to Congress in the hearings over this bill: "The satellite system is the most revolutionary communications development in my more than 50 years in the business."

Naturally, America's twin communications cartels—AT&T in telephone service, local stations in the TV industry—viewed satellites' potential very differently. For Bud Rogers and his fellow local station owners, satellites threatened to bypass the distribution bottleneck that was the source of their power and profits. The leaders of the Bell System disliked the idea of losing the $50 million in annual radio and television distribution fees, but what really terrified them was the idea of competitors invading their $3.5 billion-a-year long-distance business by bouncing phone calls off satellites.

In 1962 AT&T had grown into a corporate giant with a financial might unparalleled in history. The phone monopoly owned every last piece of its communications network, from the 64 million phones its customers used every day to the 318 million miles of wire that connected them together. AT&T's network, which the company's accountants valued at $24 billion, was the world's most valuable corporate asset. The phone giant was the world's largest private employer, with 781,000 workers. (General Motors with 553,000 employees, ranked second. RCA, with 90,000, remained a relative pipsqueak.) The phone company's business was even recession-proof, having expanded for the previous thirty-nine straight years.

Since World War II, the Bell cartel had drifted away from Theodore Vail's original philosophy, which tied the right to sizable profits for AT&T's shareholders to steady improvements in the cost and quality of its phone service for customers. Since the end of the war, the phone monopoly had abandoned the old Vail tradition of cutting the price per call (and making up the difference by carrying more calls) and replaced by a more shortsighted view. Since 1945, the number of a local phone calls was up 150 percent (from 91 million a day to 231 million), while the total tab Americans paid for local service rose over 350 percent (from $1.1 billion to $5.1 billion). That meant the average cost of local phone service jumped from three cents a minute in 1945 to six cents a minute in 1962, despite the enormous cost reductions made possible by microwaves and other technological advances. Inflation accounted for one penny of that increase. The other two pennies went into the pockets of AT&T's executives, its well-paid workforce, and its legions of shareholders.

AT&T built its network with money tapped from millions of investors large and small. Since World War II, the company had raised $9.2 billion by selling new shares of stock, almost a quarter of all new stock issued in the United States during the period. Shares of AT&T were America's favorite investment. The company's two-millionth shareholder, an all-American family from Wichita, Kansas, had recently traveled to New York so AT&T chairman Frederick Kappel could present them with their stock certificates. (The Poulsons had smiled broadly and declared their intent to use the investment to send their two children, Kay and Steve, to college.) AT&T lobbyists rarely failed to mention how many of a congressman's constituents were also AT&T shareholders.

The satellite brawl in Washington began when the Kennedy administration proposed flexible rules for the new industry, along with explicit provisions designed to prevent AT&T from extending its monopoly into orbit. AT&T and its broadcaster allies responded by backing a competing bill written by Oklahoma senator Robert Kerr, who also professed a desire to make sure the new industry got off to the best possible start.

According to Senator Kerr, President Kennedy's plan to allow unfettered competition in space posed the real danger to the new industry. If AT&T, RCA, Westinghouse, and other big companies all launched satellites, the senator argued, soon competition would be so fierce that none of them would make money and the industry would wither. To help these corporations avoid investing their money so foolishly, Congress needed to create a single company that would be responsible for running a single satellite communications network, Senator Kerr said. Under his plan, AT&T would be the new corporation's dominant shareholder, but the FCC would monitor it closely to make sure the phone company lived up to its promise to let competitors use the satellites on a fair and equal basis.

The phone giant's executives, led by Vice President James Dingman, had been roaming Capitol Hill for months, extolling the virtues of Senator Kerr's approach. Dingman told every legislator he could that launching satellites would be incredibly expensive, and so there was no need for a bunch of different companies that duplicated one another's efforts—reprising the original justification for creating AT&T at the turn of the century. Ma Bell was simply interested in helping the promising technology develop to its full potential, Dingman promised. "This position may be construed by some as stemming from the selfish interests of my company, which is the largest of the carriers involved," he testified. "Let me assure you it is not. Let one thing be crystal clear: AT&T has no desire or intention of seeking to control the communications satellite system to its own competitive advantage."

At first, the Kennedy administration fought back against Kerr's bill. Lee Loevinger, who ran the antitrust arm of Robert Kennedy's Justice Department, scoffed at the problem the bill purported to fix and suggested Ma Bell's interest in adopting the infant industry was less maternal than filicidal. AT&T's power came from its land-based network of metal wires, Loevinger told the Senate's monopoly subcommittee. Putting it in charge of space-based satellites that one day might render the land-based network obsolete, he testified, made as much sense as responding to the invention of the automobile with a decree that all motor transportation must be owned by the railroads.

David Sarnoff, who also testified before Senator Kefauver's subcommittee, pointed out that government policy had done a fine job of introducing competition into the business of moving telegrams and other digital messages—most everything except phone calls, which remained the exclusive province of the phone company. The General argued that if Congress were to establish a regulated satellite monopoly, it had to make sure that the company brought competition to the movement of all kinds of information—including phone calls.

Such concerns hadn't seemed to matter. Once again, the Washington machinery that had proven so effective at crushing new communications technologies revved up. As the competing bills made their way through the legislative process, AT&T's lobbying power slowly pushed aside congressional worries about its intentions. Eventually, President Kennedy agreed to support the bill that the senators were now debating.

As Johnson listened impassively, many of the filibustering lawmakers used their time to accuse Kennedy of having caved to the phone company's pressure without even having the decency to admit that he had abandoned the fight. Louisiana senator Russell Long hit this point with his usual dry wit: "There were two bills introduced to Congress, the Kennedy Bill and the Kerr Bill, both authored by distinguished friends of mine," Long drawled. "And I must say, as I look at the progeny, it looks a little more like my friend Senator Kerr than it does look like President Kennedy."

As for his colleagues' assurances that the Federal Communications Commission would oversee AT&T's new satellite monopoly to make sure it didn't abuse its power, Long quipped: "If the FCC cannot control AT&T on land, how in the world can it do it in outer space?"

To Lyndon Johnson, the objections of the filibustering senators sounded both sanctimonious and naive. To his way of thinking, turning the public airwaves into private wealth offered an excellent way for clever politicians to get rich. During his twenty years in Washington, Lyndon Johnson had settled on a single rule: be discreet about the graft.

IN 1942, A few years after arriving in Washington as a freshman congressman, Johnson learned his first lesson in how *not* to run the airwave con from a Georgia congressman named Eugene

"Goober" Cox. Immediately after lobbying the FCC to grant an AM radio license to Albany Herald Broadcasting, Congressman Cox had cashed a $2,500 check from the broadcaster in payment for unspecified "legal services." The same company then sold Representative Cox $2,500 in stock at prices that seemed suspiciously low. When FCC staffers discovered the transactions, FCC Chairman Lawrence Fly made the obvious—and in retrospect terribly naïve—decision to forward the evidence to House Speaker Sam Rayburn for a possible investigation into influence peddling.

Eager to squash the scandal, Rayburn called in the Georgia congressman and offered to sweep the accusations under the rug. That wasn't good enough for Goober Cox. His honor had been insulted, and he demanded the insolent regulators be made to pay. The FCC was "the nastiest nest of rats in the country," he fumed, and had "established terroristic control of all media of communications." Capitalizing on a widespread sense in Congress that the eight-year-old agency was a failed experiment, Cox requested a congressional investigation into the commission, and soon after the House of Representatives voted to appropriate $60,000 to create a special subcommittee to do just that, installing none other than Goober Cox to lead the investigation.

Chairman Cox dispatched private investigators to dig up dirt on members of the FCC, especially it's chairman, Lawrence Fly, who had a reputation as a playboy that dated back to his days running the Tennessee Valley Authority. After one of the investigators turned up evidence of an extramarital affair from Fly's TVA days, Representative Cox used it to pressure the FCC Chairman into backing off his original charges against the congressman.

The story might have ended there, if not for the bit part that Lyndon Johnson played in the scandal. Clifford Durr, one of Chairman Fly's allies on the FCC, happened to be a neighbor of Johnson's. Then-congressman Johnson suggested Durr leak the story to the *Washington Post.* A few days later, the paper ran a front-page story laying out the details of Congressman Cox's behavior (minus his investigation of Chairman Fly's affair). The *Post* also ran a front-page editorial that excoriated Cox's investigation of the FCC as "a mockery of justice [that] gives narrow personal prejudices and interests precedence over the public welfare."

Lyndon Johnson had his own reasons for wanting to help defend the FCC from the Cox investigation. While Cox and Fly were busy squabbling in Washington, Johnson was looking into buying a Texas radio station, focusing his attention on a struggling one in Austin, KTBC. The AM radio industry of the early 1940s was very profitable as a whole, generating pretax profits of $45 million on $190 million in revenue, but a few radio stations continued to lose money year after year. KTBC was one of them. The cause of the station's financial woes was obvious. The FCC had sandwiched the small station between two higher-power San Antonio stations near the top of the AM dial. Even worse, the commission only allowed KTBC to broadcast during daylight hours. At night, KTBC had to cede the frequency to a radio station run by Texas A&M University.

In the lingo of the industry, KTBC was a "sundowner," and experienced broadcasters dreaded owning a sundowner. Being confined to daytime-only hours amounted to a virtual death sentence. The most popular radio programs were broadcast at night. AM radio waves didn't travel as far during the day, when sunshine limits the waves' ability to bounce off the ionosphere. To top it off, sundowners rarely managed to land affiliations with CBS, NBC, and ABC, which preferred to sign up twenty-four-hour stations.

KTBC's financial records testified to its troubles. In 1942, when LBJ began eyeing the station, it was on track to lose $6,000, bringing its accumulated debt to nearly $25,000. The increasingly desperate owners had been begging the FCC to save the station by moving it to the uncrowded end of the radio dial and letting it broadcast at night. Despite the owners' extensive engineering reports on the wisdom of the idea—and the obvious benefits to the Austin-area listening public in having an additional nighttime radio station on their dials—the FCC repeatedly refused.

That changed as soon as Lyndon Johnson showed up. Using $25,000 of his wife's money, which she had recently inherited, Johnson negotiated a deal for her to buy KTBC. The price was a key part of the plan. Paying $25,000 for a money-losing sundowner was certainly fair, perhaps even generous. What's more, the purchase also appeared to be a real risk for the Johnsons, who sank nearly a third of their wealth into a money-losing business. Lady Bird Johnson filed an application with the FCC to purchase KTBC on January 23, 1943—four days after Congressman Cox announced that the House of Representatives would be investigating the FCC. The commission approved the transfer request to the wife of their secret Congressional ally in less than three weeks.

Four months after that, as Chairman Fly fought to defend the FCC from Goober Cox's investigation, Johnson began the next phase of his plan. Lady Bird petitioned the FCC to let her station move to the less-crowded end of the dial. She also asked for the license to be modified to turn the "sundowner" into a full-time station. After years of saying no to identical pleas from the previous owners, the agency took less than a month to reverse its stance on both counts. Overnight, the Austin-station broadcasts could be heard in at least thirty-eight counties in the heart of Texas. When Johnson bought the station it had been selling $2,600 in ads a month. By the end of 1943, as a full-time station, that number jumped to $5,645.

To Johnson, this was not nearly enough. The lanky congressman started taking trips to New York to see Bill Paley and Frank Stanton, the men who ran CBS. Often he simply showed up in their offices unannounced. The congressman told the CBS brass that he badly wanted to affiliate with CBS's top-rated network. (The CBS affiliation, he wrote in a private letter, "is life and death to us.") Normally, such a request would have been reviewed by CBS's affiliate relations department and summarily denied. The network already had an affiliate in San Antonio whose programs played in Austin. After LBJ dropped by Bill Paley's office at CBS, however, CBS decided the conflict with its existing affiliate wasn't a problem, after all.

His CBS deal in hand, Johnson headed to the office of ABC's chief executive, Leonard Goldenson. Ostensibly, Johnson's visit was intended to get Goldenson to look over KTBC's programming and offer any suggestions on how to improve it. No fool, Goldenson took the opportunity to offer Johnson a favor, calling up J. Walter Thompson and a few other major New York ad agencies to suggest they buy ads on the young congressman's radio station. By 1944, ad sales at KTBC had more than doubled again, to $13,500 a month.

Johnson was similarly shameless in persuading Texas businesses to become advertisers. The list included grocery stores and other local companies, which had a legitimate interest in reaching his listeners, as well as many companies whose main business was selling to the U.S. government, not the citizens of Austin, Texas. To anyone familiar with this side of LBJ's business dealings, most of which remained private, the implication was obvious. Lyndon Johnson wasn't really selling ads. He was selling influence.

Though Lawrence Fly's chairmanship was set to run through 1949, after his battle with Chairman Cox he decided to retire almost five years early. That proved no problem for Lyndon Johnson, who quickly struck up a friendship with Fly's replacement, former CBS lawyer Paul Porter. In 1945, Lady Bird Johnson asked the FCC for permission to quintuple the power of KTBC's broadcast signal, expanding the station's reach to another twenty-five counties. Porter's commission quickly agreed that such a move would be in the public interest. By 1946, KTBC's ad sales had nearly doubled yet again, to $22,000 a month.

Through it all, Lyndon Johnson stuck to a simple story: he had no involvement with the station, which was owned by his wife. He credited the station's success to Lady Bird's hard work and business acumen. He also made a particular point of noting that by law, her assets from her inheritance remained solely her property, not his: "I don't have any interest in government-regulated industries of any kind and I never have. All of that is owned by Mrs. Johnson."

While this was technically true, all the profits that flowed out of the station landed in the Johnsons' joint bank account. By the 1950s, that flow had become a flood. Lady Bird drew a

$21,500 annual salary, more than her husband's congressional paycheck. Far more importantly, she could siphon cash from the station's rapidly growing bank account whenever she felt like it, or whenever her husband asked.

Over the years, as Congressman Johnson became Senator Johnson and his net worth swelled, he watched new generations of lawmakers forget the lesson of Goober Cox, failing to take the same precautions as Johnson had to disguise their graft. In 1958, a law professor named Bernard Schwartz exposed Representative Oren Harris's sweetheart deal to buy 25 percent of an Arkansas VHF station for just $5,000. An artless con man, Harris had been far too blatant in his version of the scam. The station already had $100,000 in hard assets, so even ignoring the value of the FCC license, which was worth many times that amount, paying $5,000 for a 25 percent stake was the height of grabby incompetence. His business partner's paper-thin cover story—I sold it to the congressman cheap because we were childhood pals—sealed Harris's fate. While the congressman escaped any formal punishment, the publicity forced him to abandon his investment. A few years later, the remaining owners of the station flipped it for $1.1 million, of which Harris got nothing. Had he been smart enough to put up a reasonable amount of cash to purchase his stake, he would have walked away a rich man.

Other lawmakers had done a better job of running the scam. No one made a fuss over the holdings of lawmakers like Senator Clinton Anderson, a Democrat from New Mexico, who had acquired, under somewhat mysterious circumstances, a 5.5 percent stake in his hometown TV station worth roughly $152,000, the equivalent of six years of senatorial paychecks. While no one kept a definitive list of how many lawmakers had gotten in on the game, when Johnson was majority leader that group included both himself and the Republican minority leader, William Knowland, co-owner of an Oakland radio station. In both the House and Senate, the committee charged with overseeing the FCC was run by a member of the station-owning clique—or had been until the bumbling Representative Harris had been forced to sell. In the Senate, the job of running the Commerce Committee remained in the hands of Warren Magnuson, part owner of Seattle radio station KIRO.

When the FCC ended its four-year study of how to promote competition in television in 1952, the opportunities for lawmakers to get rich off the airwaves instantly multiplied. While the FCC focused its public efforts on licensing a huge number of UHF stations that almost no one would ever watch, it also decided to squeeze in a few more VHF stations in a handful of cities. Several lawmakers tried to profit from that opportunity in the established way, quietly arranging for small ownership stakes in exchange for their influence with the FCC.

When the FCC found room to squeeze in a single additional VHF station in Seattle—by that point a guaranteed ticket to tens of millions in profit—it happened to go to the radio station partially owned by Senator Magnuson. When his opponent in the next Senate race cried foul, accusing him of a conflict of interest—chairing the committee that oversaw the FCC while simultaneously profiting from the agency's gift of a VHF television license—Magnuson declared it wholly unfair to accuse him of a crime without concrete proof. "The wickedest nonsense would be his assertion that the Eisenhower-appointed FCC is subject to influence," said Magnuson's indignant campaign spokesman. The issue never captured voters' interest and the senator cruised to reelection, demonstrating once again the advantages of using an invisible resource that voters couldn't see and didn't understand as a currency for a political payoff.

Despite his many imitators, no lawmaker had ever come close to matching the scale of Lyndon Johnson's broadcasting ambitions. Austin was an obvious candidate for a new VHF station in 1952, since its residents had waited out the FCC's four-year license freeze without any local stations at all. Everywhere else in the country, competition for any of the handful of new VHF licenses was fierce. Well-financed applicants hired consultants, lobbyists, engineers to put together elaborate applications in anticipation of having to duke it out with other applicants in front of the FCC. In Austin, however, after Lady Bird Johnson applied, no one else bothered. Why enter a fight that everyone knew LBJ was going to win?

Every decision the FCC made regarding the Johnsons' new TV station seemed to break in the senator's favor, just as it had for him in the radio business. In smaller Texas cities like Corpus Christi (with 75 percent of Austin's population) and Amarillo (with 60 percent), the FCC found room for three TV stations. After granting Johnson Austin's first TV station, however, the commission had decided that city didn't need any more. As a result, Johnson managed to do his Austin radio station one better, using his monopoly to win affiliations with all three networks, allowing him to broadcast whichever network's show would attract the highest ratings.

Johnson even found a way to make money from UHF television, which, by the late 1950s, had been exposed as the disaster Edwin Armstrong and others had warned the FCC it would be. Acting through his wife, Johnson worked out a deal to buy a

UHF TV station in Waco, Texas, for $25,000 in cash plus another $109,000 in assumed debt. The sale price, far from looking like a sweetheart deal, actually appeared a bit generous. The station was on channel 34, a UHF station at a time when less than 10 percent of the sets had dials that allowed them to tune in UHF stations. Adding to its problems, channel 34 was an affiliate of the still-struggling ABC network, while a competing VHF station in Waco was set to become an affiliate of the top-rated network, CBS. (NBC already had an affiliate in nearby Temple, Texas, and planned to honor that deal.) After Johnson's purchase, however, CBS abandoned its affiliation plans with the Waco VHF station, KWTX, and instead signed Johnson's UHF station. All of a sudden the city's clearly inferior TV channel was its sole CBS affiliate.

It was the closest that Lyndon Johnson ever got to overreaching. The owners of KWTX, apoplectic at getting spurned by the networks, filed a fifty-page brief with the FCC accusing Johnson's company, ABC, and CBS of a conspiracy that violated every major antitrust law on the books. CBS's decision could not have been "motivated by ordinary business judgment," the filing pointed out. And the result—a two-station town in which only one station could air network programming—represented an obvious conflict with the FCC's stated desire to give viewers "the widest choice" in TV programs.

Adding to the anger of KWTX's owners, the FCC had chosen that very moment to allow Johnson's Austin TV station to crank up its transmitter's power, just as they had done so often with his AM radio station in the past. Waco sits one hundred miles north of Austin and with the new power, KTBC's programs could reach a large section of the southern Waco TV market.

Johnson moved quickly to protect his growing media empire and quash the feud. Eighteen days after the outraged KWTX station owners filed their challenge with the FCC, they withdrew it. Shortly thereafter, the two sides revealed the terms of their truce. In exchange for 29.05 percent of Waco's VHF station, Johnson agreed to transfer the affiliations of his UHF station, which he would then close down. Both Johnson and the station's original owners did well in the deal: the original owners kept 71 percent of a VHF station that suddenly had the Waco market all to itself, plus the benefit of Johnson's sway over the FCC the next time a regulatory decision threatened their profits. Waco's television viewers and advertisers were the obvious losers, a fact that the FCC was content to ignore.

Within two years the FCC granted LBJ's Austin TV station another major power increase, again increasing its viewership and value. The senator also negotiated the purchase of a third TV station, in Weslaco, Texas. Soon after, the commission agreed to let that station raise its transmitting power as well. In Bryan, Texas, after the Johnsons bought an educational TV station, the FCC did the senator and his wife an even more unusual favor, deciding that the public interest would best be served by allowing the Johnsons to convert it to a commercial station.

AS LYNDON JOHNSON listened to yet another filibustering senator mock the satellite bill, he tried to keep his focus on his goal. After all his efforts, he had no intention of allowing satellites to start poaching the viewers that he had worked so hard to have all to himself.

In his days as majority leader, Johnson had taken pride in entering the well of the Senate knowing just what was going to happen next. "You've got to work things out in the cloakroom, and when you've got them worked out, you can debate a little before you vote," he liked to say.

Today the vice president knew he had no chance of rounding up the sixty-seven votes that constituted two thirds of the hundred-member chamber, but he didn't intend to let that stop him. Johnson was most interested in the senators he couldn't see—they were central to his plan. Arkansas senator J. William Fulbright was nowhere to be found, he noted. Nor were four other southern senators who had refused to vote to break the filibuster—not because they opposed handing control of the satellite industry to AT&T but because they feared their votes would be used against them the next time they tried to filibuster civil rights legislation. According to Senate rules, ending a filibuster didn't take two thirds of the entire Senate, just two thirds of the senators available to vote. Along with a handful of other senators who were away for legitimate reasons, Johnson calculated that the absence of the southern five dropped the necessary number of "ayes" from sixty-seven to sixty-three.

As Johnson began to move for a vote to end debate, other veteran legislators caught on to his game—but there was nothing they could do. Senator Everett Dirksen was droll in defeat: "A lot of things can happen to a Senator on the way to the Senate. Someone might stop a Senator and try to sell him a horse. You can lose a lot of money in a horse trade."

Almost before the filibustering senators knew what had happened, Johnson had assembled what the the *New York Times* described as "an unusual assortment of Republicans, Democratic moderates and Southern Democrats" to vote to end debate. Johnson's cloture motion squeaked through, sixty-three in favor, twenty-seven opposed. After that, the dissident senators had no choice but to give up the fight.

TWO MONTHS after Congress passed the legislation to outlaw competition in the satellite industry, a Delta-B rocket took off from Cape Canaveral, Florida. A few minutes later, it dropped the RCA-built Relay-1 satellite into orbit. As an experimental satellite owned by NASA and not designed to carry commercial traffic, the satellite didn't violate the new AT&T-backed satellite monopoly. Still, the successful launch was a bittersweet moment at RCA, since the power of satellite technology would soon fall under the control of its bitter rival.

Scheduled to carry a televised message from President Kennedy to the people of Japan, the first television images to ever leap the Pacific Ocean, Relay-1 instead was pressed into service transmitting news coverage of President Kennedy's assassination, as well as footage of America's new president, Lyndon Baines Johnson.

As the new leader of a grieving nation, Johnson largely escaped scrutiny over his collection of broadcast properties. Eventually, the press did dig up pieces of the story. A *Life* magazine investigation in 1964 put the size of the Johnsons' broadcasting holdings at $8.4 million, up over three-hundred-fold from the $25,000 they had invested twenty-one years earlier. (And that didn't include the money they had taken out along the way.) As for LBJ's staunch denials of active management of

LBJ Company, the *Life* magazine report and a similar investigation by the *Wall Street Journal* found plenty of business associates willing to describe, in vivid detail, just how deep the new president's involvement had been in every corner of his broadcasting business. "There is little doubt in Austin that Lyndon is deeply interested and involved in his business enterprises thereabouts," *Life* noted, "Or that he is touchy as a new sunburn about mention of the fact."

Unfazed, the broadcaster-turned-president continued to manage his broadcasting empire from the White House. Leonard Goldenson, the head of ABC who had received Johnson's impromptu visits and fielded his brazen requests for favors for the last twenty years, was shocked at how blatant the new president could be. On one visit to the Oval Office, President Johnson complained to Goldenson about the way the press had questioned his control over Lady Bird's media holdings, despite the fact that the TV and radio stations had been put in a "blind trust" meant to avoid any conflicts of interest that might arise from a sitting president conducting private business affairs on the side. A few months later, Goldenson got a call from the president, pressuring him to move ABC's popular college football games onto his stations. "I thought about reminding him how critical he'd been of the way the Kennedys ran their blind trust," recalled Goldenson in his autobiography. "Here he was, involving himself in the day-to-day management of a station that was supposedly in his own blind trust.

"But I didn't."

A few days later, Clark Clifford, the president's close adviser, came to New York to hammer out the details of the agreement with ABC's affiliate relations department.

FOR THREE DECADES, America's communications cartels had been protected by their friends in the FCC and Congress, crippling technologies including superpower AM radio stations, FM radio stations of all sorts, television, microwave relays, and, most recently, satellites. Now, with one of their own in the White House, the local station owners had become even more powerful.

And yet, far from feeling despondent, by the time Lyndon Johnson entered the White House the ever-optimistic David Sarnoff had already shifted his attention to yet another futuristic technology, one final opportunity to ignite a rebellion that would finally bring his enemies' empires crashing down.

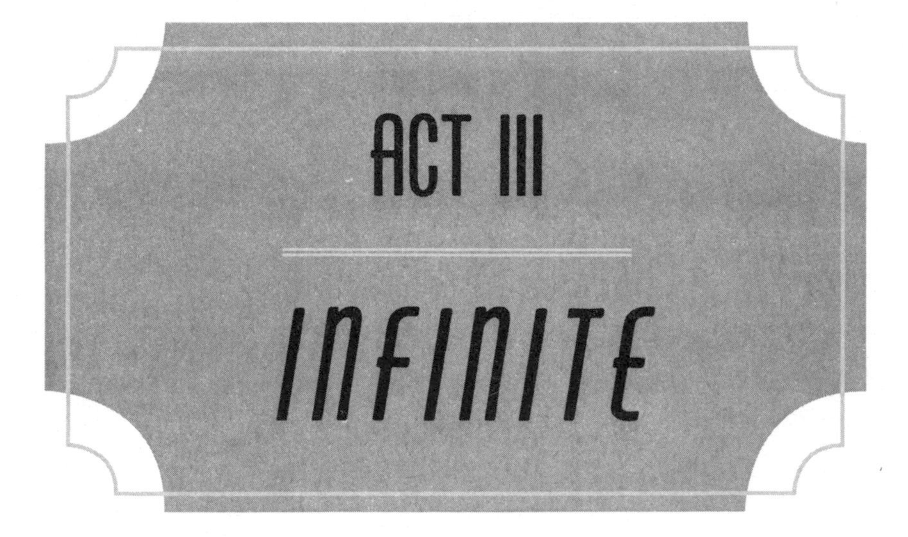

ACT III

INFINITE

CHAPTER 16

SARNOFF

New York City—1965

DAVID SARNOFF LOOKED OUT OVER THE GRAND BALLROOM AT THE Waldorf Astoria. He had been coming to the famous hotel ever since his days at the Marconi Wireless Telegraph Company, when the United States Senate panel investigating the *Titanic* disaster summoned Guglielmo Marconi to the Waldorf to testify. The celebrity-studded party that introduced the new National Broadcasting Company to the world in 1926 had featured a program of a symphonies, arias, and speeches broadcast from the Waldorf to some twelve million radio listeners. The hotel had outgrown its original home and relocated uptown to its current spot on Park Avenue in 1933, the same year that RCA moved into Rockefeller Center, just a few blocks away. Ever since then, the Waldorf had been the site of countless RCA corporate events, its steak house serving as a virtual company cafeteria for RCA and NBC's top executives. Hotel guests sometimes mistakenly assumed that the Waldorf's Peacock Bar had been named after the NBC executives who could be found drinking there.

Madison Avenue's most powerful executives filtered into the ballroom. Tonight's dinner was one of advertising's biggest bashes, an annual event hosted by the industry's nonprofit arm. Its official purpose was to honor the personal philanthropic efforts of a major corporate leader. The previous year, Charles Mortimer, chief executive and chairman of General Foods, won the honor. This year, it was David Sarnoff's turn. For most of the *Mad Men*-era advertising executives, however, the real attraction of the event was the opportunity to drink, smoke, network, and brag about the year's record billings.

Thanks largely to the technologies that David Sarnoff had spent his career popularizing, the advertising industry was living in a golden age. The decade-long boom in TV sales had sent profits soaring. This year, the advertising industry's revenue had hit $250 million from TV ads alone, along with another $80 million from radio spots. And yet, despite the debt that the advertising community owed Sarnoff, their relationship had always been chilly. General Sarnoff looked forward to socializing with advertising executives about as much as he did with actors and comedians—and the admen returned his disdain.

Much of their criticism was justified. Whatever the seventy-four-year-old's talents as technology executive, he had demonstrated little skill at running a television network or hiring people who could. Another round of management turmoil at NBC was demonstrating that shortcoming once again. The man running the network, Robert Kintner, had become known on Madison Avenue for poor programming choices and a fondness for tumblers of vodka. This past fall, stories that the free-drinking advertising executives could laugh off had given way to darker tales of Kintner's descent into alcoholism, including his out-of-control and very public drunkenness at an NBC affiliate meeting in Mexico. Naturally, much of the ad executives' predinner gossip centered on the latest management chaos at NBC, the turmoil adding to the sense that the Sarnoff era at RCA had finally reached its end.

WHILE AT&T'S RECENT SUCCESS in grounding the new satellite industry was the sort of thing that would have infuriated and obsessed a man like Major Armstrong, it no longer bothered, or even much interested, General Sarnoff. To be sure, he still believed that bouncing microwaves off orbiting spacecraft posed a mortal threat to the phone monopoly's old-fashioned network of metal wires, as well as the local broadcasting cartel's TV and radio stations. Powerful as satellite technology could be, however, Sarnoff now considered it almost insignificant compared to the power of the new type of communications that he was about to reveal to the ballroom full of advertising executives.

Charles Mortimer, last year's "Public Service Award" winner, ascended to the dais and called for the attention of his increasingly inebriated audience. Mortimer paid a brief tribute to David Sarnoff's long career in communications. Fifty-nine years had passed since the young Russian immigrant walked into the Marconi Wireless Telegraph Company and found a job as an office boy. Now here he was, at the tail end of a career no one could have dreamed possible, having played a major role in every form of communication that had been invented in the twentieth century. Sarnoff shook hands with Mortimer and took the podium, a white carnation pinned to his tuxedo's lapel, a bit of the old power back in his voice.

> "We stand on the threshold of a new era in communications," he began, "in which the physical barriers of space and time will be abolished and in which a global system of instant sight and sound will link people everywhere. It will provide communications media with the ability for the first time to reach the entire population of the earth simultaneously.
>
> "The rapidity and sweep of technological advance in the 20th century have already conditioned us to change, but let us hope that they have not made us callous toward it. For what will soon occur in communications represents change of a far different character from any that mankind has as yet experienced."

Briefly recapping the advances in microwaves and satellites, he continued:

> "We are now able to transmit across vast distances all types of information—print and picture, the spoken word, telegraphic messages, televised images, and even the esoteric language of computers.
>
> "These are only the early harbingers of tomorrow's technology. On planning boards, in research laboratories and engineering centers, further advances in electronic communications are now in development. Together they will weave a pattern of total communications, joining homes, communities and nations."

To the advertising executives in the room, General Sarnoff's grandiose prophecies of a global computer network sounded outlandish. The seventy-four-year-old charged ahead, launching into a description of how his global computer network would be made possible by the combined power of two great inventions.

The first was the microchip, which Sarnoff's engineers were now building out of silicon, one of the earth's most common elements. In the seventeen years since the invention of the transistor made Armstrong's original amplifier obsolete, scientists had learned how to control electromagnetic waves with solid chunks of silicon rather than in fragile tubes of glass. That advance had led to a handful of relatively minor breakthroughs, such as the transistor radio. Now, Sarnoff said, another technical breakthrough had made it possible to etch thousands, or even millions, of transistors into a single sliver of silicon. These microchips, he explained, could perform the same functions that used to require a complicated array of tubes, transistors, resistors, and other circuitry. Yet despite the microchip's limitless potential, so far the general public had reacted with indifference. (In one recent RCA publication, an engineer expressed bafflement that the general public had welcomed the birth of the microchip "with about as little excitement as an announcement from Detroit of a new model change.")

If talk of microscopic chips seemed fanciful to Sarnoff's nontechnical audience, the second great breakthrough that he predicted sounded loonier still. The world's networks would soon be replaced by a new type of cable, he said, made not from copper but from thin strands of glass. Inside those tiny glass fibers, laser beams would blast unimaginable amounts of information around the world.

Rumors of David Sarnoff's declining mental abilities had become a common source of gossip in the advertising community—and that was before the septuagenarian started going on about "laser pipes." Hoping to overcome his audience's skepticism, the General prodded them to learn the same lesson he had learned in his fifty-nine years of watching the back-and-forth battle between waves in the air and wires on the earth: it was only a matter of time before new technology would make the cost of moving information plunge, creating new ways for human beings to connect and new ways for savvy businessmen to profit.

Again and again, he told them, he had seen new inventions revolutionize the communications business. Mankind's ability to harness the information-moving abilities of electromagnetic waves had grown exponentially during his long career. In the 1910s, the Marconi Company sent out waves whose crests were several thousand meters from each other. In the 1920s, Edwin Armstrong found a way to use waves that vibrate much more rapidly, leaving only two hundred meters between crests. By the 1930s, Armstrong's FM radio could take advantage of ten-meter waves. Then came World War II's advances in radar research, and soon rapidly vibrating microwaves, less than a meter in length, were being used to shoot information hundreds of miles. Why should anyone expect that march of progress to stop? Laser light, after all, is merely an even shorter electromagnetic wave.

Sarnoff continued, "Laser pipes [will become] superhighways for communication among major population centers. These narrow light beams will have a total capacity millions of times greater than the most advanced systems today. Through lasers, an individual will be able to have his private 'line' for sight and sound communication across any distance, just as he now has his private telephone line."

David Sarnoff cautioned his audience against underestimating the power of microchips and laser pipes. Among other things, the two technologies would lead to the creation of "continental and global networks of computer centers. These will serve scholars, scientists, professional men and businessmen as instant sources of all known and recorded data on any conceivable subject, from ancient history to market trends, from social statistics to medical knowledge. Already, computers have been linked experimentally across the Atlantic. In due time they will communicate freely with each other, as well as with people, regardless of the distance involved."

Computers linked by laser pipes (a technology that future generations would come to know as fiber-optic cables) would eventually learn to handle every sort of communication, Sarnoff continued. At some point, all information would be digital, he said, broken down into individual bits before being transmitted from one place to another. The twentieth-century tradition of building separate networks for separate applications—a phone network for phone calls, a cable-TV network for TV shows—would give way to a new world dominated by a single digital network.

"There is no longer any distinction among the various forms of communications. All of them—voice or picture, telegraph or data—pass simultaneously through the same relays in the form of identical electronic pulses. Henceforth—in marked contrast to the past—developments that will extend the reach of one will extend the reach of all.

"This same process of unification will inevitably occur, I

believe, in all media of communications. Not only television and telephone but books, magazines and newspapers will be converted into identical bits of of energy for transmission over any distance. At the receiving end, these electronic signals will be converted into any form we choose—in visual display or recorded sounds or printed pages.

"A single integrated system means that the major channel of news, information, and entertainment in the home will combine all of the separate electronic instruments and printed means of communication used today—television set, radio, newspaper, magazine and book.

"The home will thus be joined to a new, all embracing informational medium with global reach. This medium will serve a vast public of differing nationalities, languages and customs, and its impact will be profound."

The General wrapped up his speech with some thoughts on the potential of global communications to increase understanding among nations.

As applause filled the ballroom, it was difficult to tell how much, if any, represented enthusiasm for Sarnoff's radical prophecy, how much reflected respect for the old man's legacy, and how much simply signaled relief that he was done talking.

IN THE DAYS and weeks following his speech, the public, like the audience at the Waldorf, greeted Sarnoff's grand prophecy with indifference. He ordered large excerpts of his speech to be published in various RCA publications. With minor exception, the press ignored it. The very things that would make David Sarnoff's final vision so astounding to future generations consigned it to obscurity in 1965. Describing the Internet in vivid detail hadn't cemented David Sarnoff's reputation as the most gifted technology oracle of the twentieth century. It merely made people think he was a doddering fool, talking through his hat.

The admen in the room had no doubt where the real power lay in the broadcasting industry. The Association for Maximum Service Television liked things just the way they were. The American Telephone and Telegraph company liked things just they way they were. So did Congress, the Federal Communications Commission, and Madison Avenue. And yet here was David Sarnoff, babbling on about an imaginary global computer network that would lay all of those mighty powers to waste.

CHAPTER 17

PALEY

New York City—1971

WILLIAM PALEY WALKED SLOWLY, THE WEIGHT OF DAVID SARnoff's casket pressing down on his shoulder. The synagogue on Fifth Avenue was New York's largest and, in the opinion of many, one the world's most beautiful as well. Designed in 1927, the building borrowed architectural elements from Romanesque, Byzantine, Moorish, and Gothic synagogues. At the same time, a hidden steel skeleton made its vast central sanctuary unlike any other, free of the massive interior pillars used to support the roof in older synagogues. Visitors' eyes were immediately drawn to the gilded ceiling 103 feet overhead, which was decorated with a vivid pattern of red, blue, and gold. They also tended to catch on the endless number of details that had been lovingly crafted by New York's finest Jewish artisans. In Temple El-Emanuel, even the radiator grilles were elaborate works of art.

As Paley and his fellow pallbearers made their way toward the pulpit, the assembled mourners rose to their feet. Some seven hundred of the nation's top politicians, celebrities, scientists, judges, corporate leaders, and military officers had come to pay their respects to Brigadier General David Sarnoff. High above, the midmorning sunlight streamed in through the synagogue's stained-glass windows.

Bill Paley knew that his selection as a pallbearer would surprise many of the mourners. The other men helping carry the casket were all old friends of Sarnoff: the former head of the U.S. Army Signal Corps (Major General Harry Ingles), New York's senior senator (Jacob Javits), the former head of RCA Laboratories (Elmer Engstrom), two top Wall Street financiers (Paul Mazur of Lehman Brothers and André Meyer of Lazard Frères), and the rabbi who ran the Jewish Theological Seminary (Louis Finkelstein). As the controlling shareholder and chief executive of CBS, Bill Paley was better known as one of David Sarnoff's fiercest rivals. Reaching the front of the temple, the pallbearers slowly lowered the casket near a small lectern. Paley took his seat as the governor of New York rose to give David Sarnoff's eulogy.

THAT HIS RELATIONSHIP with the General had thawed probably surprised Bill Paley as much as it had anyone else. For the last three years, Sarnoff's declining health had forced him to abandon his plans to remain at RCA and oversee the scientists building faster laser pipes and more powerful microchips. Instead he had spent his days in a series of hospitals, discussing increasingly grim medical options with his doctors. After one surgery failed to clear an infection in his mastoid bone near the base of his skull, surgeons tried twice more. Each effort left Sarnoff increasingly incapacitated and deformed.

As the General faded from the public eye, his invitations to Bill Paley to visit became more frequent. Paley often stopped by the solarium of Sarnoff's Upper East Side town house, which had been converted into a private hospital room, for quiet talks that seemed to distract Sarnoff from his suffering. In their younger days, the two men had often crossed paths at the "21," a famous restaurant both frequented. At Sarnoff's sickbed, the restaurant delivered their favorite meals.

To keep her husband entertained, Sarnoff's wife, Lizette, had moved his old telegraph key from his desk in Rockefeller Center to his bedside. When Sarnoff developed difficulty speaking, he tapped away in Morse code to old colleagues at RCA Global Communications, who had their own keys at the ready whenever their old boss felt like a digital chat. In recent weeks, as Sarnoff's condition worsened, however, Paley visited less and less often. Eventually, the General even stopped using his old telegraph key. Once the old wireless operator's fist gave out, his heart soon followed.

NBC interrupted its Sunday-morning broadcast of *Meet the Press* with the news of Sarnoff's death, and soon various tributes to the General's epic career filled the country's airwaves and front pages. NBC aired an adulatory and prepackaged special celebrating its longtime leader's life. Even less-biased obituaries marveled at the accomplishments of the man who had sailed into New York Harbor as a nine-year-old boy and gone on to foresee every major communications advance from the wireless telegraph to satellites—and fought to bring them all to the general public.

"His knowledge and ambition were the driving force behind the development of the electronic media and their profound effect on American life," wrote Jack Gould, the *New York Times'* television and radio critic who had covered Sarnoff for decades.

Of all the tributes, surely none would have meant more to Sarnoff than the impromptu memorial at the David Sarnoff Research Center in Princeton. There, RCA researchers assembled unbidden to honor the man who had done so much to support their work. Several of the scientists—*his* scientists, as the General always preferred to call them—brushed away tears.

Standing under a wooden pulpit canopy that helped to amplify his voice, Governor Rockefeller retold the now familiar story of David Sarnoff's rise from the slums of the Lower East Side to the heights of Rockefeller Center and the pinnacle of American capitalism. "His genius lay in his capacity to look at the same things others were looking at but to see far more," Governor Rockefeller said. "In others, the word visionary might mean a tendency to see a mirage. In David Sarnoff, the word visionary meant the capacity to see into tomorrow—and to make it work."

FROM WHERE BILL Paley sat, David Sarnoff's legacy appeared a bit more complicated than that. Paley and Sarnoff shared a few obvious traits—besides running America's other major radio and television network, Paley was also the son of Jewish immigrants—but it was their differences that had always defined their relationship.

Perhaps Paley's carefree childhood—and Sarnoff's lack of one—explained their ability to enjoy and appreciate frivolous pastimes. Paley found his audience's preference for *I Love Lucy* over Luciano Pavarotti neither puzzling nor dismaying. He enjoyed popular entertainment—the long-running CBS western *Gunsmoke* was his favorite—as well as the glamour that came with it. Paley delighted in hobnobbing with his famous employees, especially beautiful actresses, and he understood the power they had over audiences. He could never comprehend Sarnoff's refusal to admit the stars' strategic value or his unwillingness to pay them a salary reflecting that value. Time and again, Sarnoff's disdain for popular content led him to make mistakes. "The General sometimes demonstrated almost a genius at picking the wrong people to run a network," was the way Paley once put it.

Paley witnessed another of Sarnoff's shortcomings firsthand during the General's final years. Among his few regrets, David Sarnoff admitted to Paley, he now counted his lifelong preference for spending weekends at work instead of with his family. Perhaps in response, the General had committed himself to installing his son Robert as his successor at RCA.

That single-minded focus also befuddled Paley, given the mountain of evidence that Sarnoff's son was poorly suited to the task. After being named president of RCA in 1965, Robert Sarnoff abandoned his father's focus on technology and began reinventing RCA as a very different kind of company. Seizing on a popular management fad of the late 1960s, which held that industrial conglomerates could prosper by expanding willy-nilly into all manner of different business, the younger Sarnoff had overseen a wild acquisition spree. RCA had purchased Hertz, the country's largest car rental operation; the F. M. Stamper Company, a leading producer of frozen foods; and Coronet Industries, a carpet manufacturer. On Wall Street, the running joke held that RCA now stood for Rugs, Chickens & Automobiles.

The elder Sarnoff had made a few of his own ill-considered acquisitions over the years, but those failures were always tied to his clear vision of what RCA was: a technology company. The General's final deal, the purchase of Random House in 1966, offered a fitting example of a failed David Sarnoff takeover. Foreseeing the day when books would be delivered and read electronically, a small part of the vision he described in his speech to the Ad Council, he convinced RCA's board that purchasing the publisher would give RCA a jump on the coming market in e-books. That was the way the optimist's prophecies seemed to turn out: eerily accurate, wildly premature.

After the elder Sarnoff retired in 1966, Robert Sarnoff had allowed RCA's once sacred research budgets to wither. Three months before the General's death, he had even abandoned his father's attempt to make RCA a leader in the new computer industry. When Lizette and his nurses heard that RCA had officially exited the computer business, they made an effort to keep it from the General. He happened to hear the news while listening to the radio. "A terrible tragedy," he muttered to his nurse.

Despite those criticisms, Bill Paley also felt a deep admiration for the skills that David Sarnoff possessed and he did not. Sarnoff's biggest triumph over CBS came when he forced the FCC to reverse its selection of CBS's inferior color television technology, leaving Bill Paley with a $60 million write-down and a lesson in the perils of ignoring David Sarnoff's vision of technology's future. In a 1966 speech honoring Sarnoff's retirement, Paley summarized his admiration for his rival's technical prowess: "To all of us, David will always be broadcasting's man of the future and its most imaginative prophet."

WHEN DAVID SARNOFF'S own mentor, Owen Young, decided to retire from General Electric and RCA in 1927, his disciple set himself to the task of writing a synopsis of Young's career. How best to judge Young's many accomplishments: running General Electric, counseling presidents, teaming up with Sarnoff to create RCA? After much thought, Sarnoff began his public farewell to his beloved mentor this way: "It is said that the true significance of a man is to be found in his works. Perhaps—but not wholly. The years destroy the most enduring monuments. Circumstances often vitiate the best laid plans. Only time can tell how soundly we build.

"For the occasion this evening I should like to amend the old law. I should like to say that in his higher values a man can not be measured merely by his works; he must be measured by the things he works for; by the courage with which he faces his tasks; by the purpose that moves him; by the principles that sway his decisions."

Four decades years later, David Sarnoff returned to the Lotos Club for his retirement dinner and a chance to reflect on the meaning of his own career. After thanking his hosts and marveling at the pace of change that had taken him from Marconi's early spark-gap transmitters to RCA's satellites and microchips, he turned is attention to his favorite topic.

"I have seldom been interested in my lifetime in yesterday, and have gotten very little excitement out of today but have always been fascinated by tomorrow.

"The challenge of tomorrow is the greatest reason for our existence, and if we seek an explanation of the purpose of life it has always seemed to me that the answer is: we are here to express the forces within us, whatever those forces may be, whether they be that of a good shoemaker, of a carpenter, a good wireless operator or an engineer, or whatever else you will. If we do the best we can to express the forces with which the Lord endowed us then I think we have reason to believe that our life has been purposeful, that it has achieved such fulfillment as man can hope for on this earth."

Measured by Sarnoff's own standards, even Bill Paley would have to judge his complicated friend a remarkable success. David Sarnoff remained true to his faith in the future. The unswerving purpose that he worked for—from his first day at the Marconi Wireless Telegraph Company of America to his last day in Rockefeller Center—had been to find new ways to connect the world. When David Sarnoff dreamed, he dreamed of new ways to send information zipping around the globe at the speed of light, new ways to tie the human race together.

And, although neither the dying David Sarnoff nor his complacent, seemingly triumphant enemies had any way of knowing it, his final prophecy—a digital revolution fueled by microchips and laser pipes—had already begun.

EPILOGUE

McGOWAN

AS A KID GROWING UP IN A RURAL PENNSYLVANIA COAL COUNtry in the 1930s and 1940s, Bill McGowan never dreamed of a career as a businessman, unaware that such a profession even existed. The son of a railroad engineer and a schoolteacher, McGowan got his first glimpse of the wider world during a three-year stint in the U.S. Army in postwar Europe, after which he returned home to complete an undergraduate degree in chemistry at King's College in Wilkes-Barre, Pennsylvania. McGowan excelled at chemistry, thanks to his talent for comprehending the rules of complex systems, but found little joy in the subject. His plans for a career in medicine left him similarly lukewarm. One King's College professor surmised the gregarious, hyper-analytical student's true calling and suggested he apply for a seat in Harvard Business School's class of 1954.

Arriving on the Ivy League campus, the working-class, Irish

Catholic kid felt out of place among his patrician classmates. That changed quickly, as McGowan emerged as a star student. Harvard Business School's famous system of teaching every class through individual case studies, each one focused on a real business with real business problems, proved a perfect fit with McGowan's peculiar mind. Here were complex systems run according to complicated sets of rules—accounting rules, economic rules, legal rules—which ordinary people found mind-numbing but McGowan delighted in decoding. He tore apart cases involving individual businesses, large industries, and government agencies with ease. Once he deciphered a system's rules, he showed an equally impressive knack for thinking up ways to manipulate those rules to serve his own purposes. Show Bill McGowan a system, his classmates quickly discovered, and he'd show you a way to exploit its weakness. By the end of his first year he'd earned top marks in every class, a merit scholarship to pay for the rest of his studies, and a feeling that nothing was impossible.

After graduation, another defining characteristic of Bill McGowan's personality revealed itself. He loathed large organizations. While his old Harvard classmates sprinted up corporate ladders, a summer job with Shell Oil convinced McGowan that he never wanted to work for a large, rule-bound company again. Instead, he set up shop as a small-business crisis consultant, parachuting in to rescue one desperate firm after another. He helped save businesses that made movie cameras, suitcases, auto parts, and test equipment for NASA. To McGowan, companies were puzzles. Some puzzles had simple solutions—to revive one struggling manufacturer, McGowan turned its jewelry-cleaning device into a tool for detecting fuel leaks on navy submarines. Others required a more complicated fix.

At the top of that list was a tiny company run by a radio repairman in Joliet, Illinois, which McGowan was called in to rescue in 1968. This time, saving the small company would require destroying the largest and most powerful monopoly in American history

Every puzzle held clues to its solution, Bill McGowan believed. You just had to know where to look.

IN THE YEARS following David Sarnoff's death, the outcome of his life's battle seemed settled. The war to rule global communications was over; the cartels had won. In most countries around the world, the waves and wires connecting ordinary people were controlled by explicit government-run monopolies. In the United States, private cartels protected by a compliant government filled the same role. Everywhere around the globe, the monopoly mind-set ruled. Communications networks were not operated to maximize the ability of ordinary people to connect with each other, but to maximize the wealth their operators could extract by keeping connections scarce and prices high.

The smartest technical minds of the twentieth century had proven unable to break that control. In the U.S., challenge to the cartels' power ended the same way: another disillusioned entrepreneur leaving Washington, another technological advance regulated or legislated into irrelevance. The world's communications networks bore no resemblance to David Sarnoff's final prophecy—and there was no sign they ever would. To the extent that anyone remembered David Sarnoff's vision of a single, all-purpose global communications network, it was easy to dismiss it as a grand techno-utopian fantasy.

Bill McGowan first stumbled into the communications business after a lawyer he knew in Chicago introduced him to Jack Goeken, the radio repairman who ran the struggling company in Joliet. Goeken had founded his business after hearing truckers express frustration about being out of contact with their home offices while they were on the road. Thanks to a brief stint in the army, Goeken knew that modern microwave towers could move thousands of phone calls at a time. He proposed building a chain of microwave towers from Chicago and St. Louis that would allow truckers to call into headquarters by using their CB radio to link to the nearest microwave tower, which would then relay the conversation to either of the two cities. Goeken had asked the FCC to authorize his plan five years ago, beginning a seemingly interminable licensing process that had brought the small-time radio repairman to the edge of bankruptcy by the time McGowan showed up to help.

McGowan reviewed Jack Goeken's plan and quickly determined that he was trying to solve the wrong problem. Building five microwave towers along Highway 55 to connect Chicago and St. Louis would cost about $2 million, he calculated, and provide enough capacity to move 2,400 simultaneous phone calls between the two cities. Using those microwaves to connect truckers to their home offices would limit the company to serving a tiny market, McGowan figured, too small to justify the expense and regulatory hassle. But what if he could take the network and use it to move ordinary long distance phone calls?

Penciling out the potential profit of this investment, Mc-

Gowan could scarcely believe the scale of the opportunity. AT&T charged $2.20 a minute to call between Chicago and St. Louis, a price it justified based on the expense of building and running its network. McGowan figured the cost estimates were inflated, but was shocked by the size of the gap between the $2.20 a minute the phone monopoly charged and the actual cost of providing the call. A $2 million microwave network, using state-of-the-art 1968 technology, could move over a million minutes of phone calls during business hours each day—enough to earn back its cost of construction in less than eight hours at AT&T's going rate of $2.20 per minute. To be sure, communications networks rarely run at full capacity, but it hardly mattered. The new network could charge half of AT&T's rate, fill 10 percent of its total capacity, and still recoup its entire cost of construction in a single month.

Entranced by the challenge of toppling the country's most powerful company, McGowan did what he always did—he began to read. "Some people read books," one of McGowan's friends once observed. "Bill reads libraries."

MCGOWAN'S RESEARCH BEGAN at the start of the twentieth century, back when Theodore Vail ran the Bell system. In those days, the wisdom of having a single national phone company was rarely questioned. The primitive technology of the day settled the debate. Building overlapping phone networks out of expensive copper wire would be phenomenally expensive, and even if the country did foot that bill, without computers to switch calls between the competing networks and keep track of the billing, interconnecting multiple networks would have created endless operational headaches. The only question, in Vail's day, was who should run the monopoly: the government or a private monopoly regulated by the government.

To strengthen his argument in favor of a regulated private company, Vail constantly pushed new technology into the phone network to cut costs, add capacity, and strengthen the political case that the Bell system deserved to keep its government-approved monopoly. It helped that Vail's case that in other countries around the world the government seized control of the country's phone network immediately, installing functionaries naturally inclined to squeeze short-term profits from the citizenry by keeping prices high and connections scarce.

Digging into the Bell system's byzantine accounting system, McGowan realized that the FCC and state regulators assigned to keep the company from abusing its monopoly had instead become enablers of its absurd prices. In theory, tying prices to how much the company spent on its network made sense in theory, since private investors who put up capital to help build the phone system would only do so in exchange for a reasonable return on their investment. In practice, it became an invitation to extravagance, giving the Bells the incentive to spend as much as possible regardless of whether that spending helped its customers.

In Theodore Vail's day, the AT&T chairman devoted much of his annual letter to shareholders boasting about the company's record of continuous price cuts and the speed with which it replaced obsolete equipment. By the end of World War II, AT&T's accountants had begun a covert campaign to do the opposite, gradually lengthening their estimate of the "useful life" of new equipment in their network. To his astonishment, McGowan discovered that this game had gone on so long that the bookkeepers often assigned new communications gear a thirty-five-year life-span. The gimmick boosted the "book value" of the network and the regulated profits AT&T could collect. It also incentivized the phone company to keep obsolete electronics for decades. After years and years of shortsighted thinking, no wonder the Bell network was larded with so much inefficient junk.

Everywhere in the vast Bell system that McGowan looked he found flab. Many AT&T managers were expected to spend a full third of their time out in the community, not working on the business but hobnobbing. In order to demonstrate the phone company's support for local charities, the managers were encouraged to offer favors such as picking up the tab for printing invitations to a charity ball. Why not? The phone company got the credit, the expense got folded discreetly into the nation's phone bills. Even the army of lobbyists that AT&T employed to kill off competition counted as a business expense, paid for by the very customers whose bills the lobbyists spent their days attempting to raise.

Unlike previous generations of telecom rebels who had headed to Washington with nothing more the good intentions and superior technology, Bill McGowan took the time to study how the FCC really worked. He quickly picked up on the commissioners' long tradition of professing heartfelt support for competition even as they passed rules that crippled it.

Most recently, in a pattern reminiscent the FCC's hobbling of FM radio, the commission had cooked up another plan to help promote competition in television, this time by halting the growth of cable television. Stubbornly unwilling to admit to the failure of UHF television as a viable competitor after fifteen years of failure, the FCC continued to act as though the inferior physics and unworkable economics of UHF stations could be overcome with a few more helpful regulations.

In 1966, the commission issued a flurry of new rules banning new cable-television systems in order to stop them from draining viewers from UHF stations. The move flabbergasted antitrust experts and angered the UHF industry, which protested that it neither wanted nor needed this kind of help, since in fact the new cable-television networks helped struggling UHF stations by delivering their programs to homes that their feeble over-the-air signals couldn't reach. Once again, the FCC insisted it knew best, and once again its rules hurt the competitors they were ostensibly meant to help. After new cable systems were banned, the collective losses of UHF stations more than doubled, from $7 million to $17 million. Meanwhile the profits of independent VHF stations rose from $287 million to $321 million.

The FCC's blend of ignorance, overconfidence, and venality—the traits that had so infuriated Edwin Armstrong—didn't aggravate Bill McGowan in the slightest. It wasn't that he approved of the commission's actions. He just retained a cynic's talent for dispassionately observing the absurdities and injustices of Washington.

As McGowan saw it, the corrupt relationship between AT&T and its ostensible regulators was rarely a matter of outright bribery. In the broadcasting industry, individuals fighting for valuable broadcast licenses made direct bribes far more likely. In the phone business, on the other hand, the

Bellheads relied on the commissioners' ignorance and self-delusion to get their way. Year after year, decade after decade, AT&T's army of lawyers and lobbyists descended on Washington to repeat Theodore Vail's argument that the phone system was a "natural monopoly" and that competition would cause more harm than good. Most FCC commissioners, McGowan told his colleagues, lacked the knowledge, brains, and temperament to argue the point. Instead, they just accepted it unquestioningly. "The Commission, just about universally, for a long, long period of time, was very anti-competition," McGowan observed, more bemused than angry. "They didn't understand. They didn't work on understanding. They just thought that there was a natural monopoly."

Another time, McGowan found himself on an airplane sitting next to a former FCC chairman who had served during the Kennedy administration. "I marveled at how little—I mean, little—he understood about the telephone industry," McGowan said. "He didn't know the people, the players, the concepts, the arguments, the structure, the proceedings, nothing, nothing, nothing."

McGowan's research into the FCC's long history as AT&T's lapdog—seemingly enough to quash anyone's hopes of cracking the phone monopoly—only heightened his interest.

The fate of one small company seemed particularly intriguing. The tale started in 1948, when AT&T lawyers asked the FCC to forbid the sale of a small rubber device that could attach to a telephone mouthpiece in order to help keep a private conversation private. Use of the "Hush-A-Phone" wouldn't affect the phone company in any way, but Bell system policy dictated that it own anything and everything that attached to its network. At first, the Hush-A-Phone seemed doomed to the fate of so many other insurgents, whose technology consumers liked but regulators wouldn't allow. Then came the twist that caught Bill McGowan's attention. Instead of closing up shop, Hush-A-Phone's owners headed to federal court to sue the FCC.

Ever since 1943, when RCA's lawyer John Cahill lost the case of NBC v. FCC in the Supreme Court, the idea of suing to overturn a dubious FCC decision had seemed like a mug's game, thanks to the sweeping discretion the Supreme Court allowed the commission in that case. Hush-A-Phone's lawyers tried anyway, arguing that the FCC's discretion only went so far. How, they demanded to know, could a device that functioned exactly like a human hand cupped around a phone's mouthpiece possibly harm the phone network? What rational purpose did banning it serve? The FCC, having outlawed the device without bothering to muster a reasonable answer to either question, sent its lawyers to court to try to convince a panel of three federal appeals court judges that Hush-A-Phone really did harm the phone network—by making it slightly harder for the person on the other end of the line to hear the Hush-A-Phone user. Even if the court didn't agree with that assessment, the FCC's attorney added, it should still defer to the FCC's technical expertise and butt out.

Judge David Bazelon had fun with that argument. The Communications Act endowed the FCC with broad decision-making powers, he acknowledged, but not so broad that the agency could impose arbitrary rules that defied common sense. No one but the two people having the phone conversation could conceivably be harmed (and could quite conceivably be helped) by using the Hush-A-Phone to keep their conversation private, Bazelon observed in his decision striking down the FCC's rule.

Furthermore, the judge continued, the case highlighted the need for a general guideline that the FCC could apply in future cases. Crafting a commonsense standard that would haunt AT&T and its government guardians for the next several decades, Judge Bazelon ordered the FCC to refrain from "unwarranted interference with the telephone subscriber's right to reasonably use his telephone in ways which are privately beneficial without being publicly detrimental."

The quirky little case, settled in 1956, caused little alarm at the Bell system, at least until three years later, when a Texan named Tom Carter decided to use the court's Hush-A-Phone ruling as precedent to force the FCC to authorize the sale of his own invention. The rancher had built the "Carterfone" to automatically connect his home phone to a short-range radio, allowing him to make and receive phone calls from anywhere on his ranch. AT&T's lawyers again ran to the FCC and demanded that the "foreign attachment" be outlawed. This time, however, the FCC could not bow to Ma Bell's demand with its usual alacrity. Forced to apply Judge Bazelon's Hush-A-Phone standard to the "Carterfone," the FCC commissioners ordered staffers to determine if the invention was "publicly detrimental" to AT&T's network. It wasn't. Since all the device did was send the sound from a walkie-talkie into the mouthpiece of a regular phone, it couldn't damage the phone system any more than a regular human voice. The commissioners soon voted to follow the Hush-A-Phone precedent and authorize the Carterfone for sale.

As he read through the court cases, Bill McGowan began to

imagine ways to stretch the precedent even further, forcing the Bell System to connect his microwave towers directly into their monopoly phone network.

HAVING VERSED HIMSELF in the ways of Washington, McGowan decided to give his sketchy plan a try. He struck a deal with Goeken to pay off the company's debt, took over as chief executive, and moved the company's headquarters to Washington, D.C. He then recruited a former FCC commissioner named Kenneth Cox as the company's lawyer and set about revising Goecken's proposal to build microwave towers linking Chicago and St. Louis.

At the FCC's public hearing, AT&T sent a few lawyers and expert witnesses to testify about the reasons for rejecting McGowan's application, but didn't take the tiny challenger seriously enough to mount an aggressive campaign. The phone company's typical strategy, McGowan knew, was to barrage the commission with reasons that the proposed new network wouldn't work, which usually gave the commission a reason to kill the idea then and there. Every time an AT&T expert gave another complicated technical explanation of why the proposed microwave network would never work, McGowan dug around in his papers and feigned surprise at being able to find the perfect rebuttal. His favorite moment came when one AT&T technical expert testified that the proposed system could never work because its towers were spaced too far apart. McGowan made sure the engineer laid out his technical case in detail before pulling out maps of AT&T's own microwave network and asking the engineer to read off the distances between its towers. The humiliated engineer had no choice but to admit that they were even farther apart than the ones he had just declared unworkable.

McGowan's real strategy, however, had nothing to do with persuading the commissioners to appreciate the merits of his position. Instead, he focused on misleading them into thinking that his network would never pose a real threat to AT&T. Whenever possible, McGowan focused the commissioners' attention on the microwaves used by his proposed network–a meaningless technical distinction that he figured would be a good way to distract them from what he really wanted to do.

McGowan's combination of preparation and prevarication worked–just barely–when the commission voted four to three to approve his five microwave towers and new status as the first "specialized common carrier."

Immediately after winning approval for his new network, McGowan began the second phase of his plan, which he also kept hidden from the commission. Traveling the country, he began to recruit groups of local investors to fund city-to-city microwave links similar to the one between Chicago and St. Louis. He soon raised enough cash to connect another sixteen city pairs with microwave links. Using Goeken's application as precedent, McGowan then began to flood the commission with copycat applications that he knew would no longer require a formal hearing. After winning those approvals, McGowan again surprised the commission by snapping the tiny companies together to form a microwave network stretching across the country.

By June of 1972, six months after David Sarnoff's death, McGowan's newly renamed company was ready to go public.

Microwave Communications Inc.—MCI for short—sold 3.3 million shares of stock and raised $30 million to continue its construction binge.

THE METEORIC RISE of MCI, from its roots as an unheralded business in rural Illinois to a publicly traded company with $30 million in the bank, spurred AT&T to serious action. The month before MCI's stock offering, the Bell system's top executives gathered in Key Largo, Florida, for their annual meeting—and the need to knock off Bill McGowan and MCI was the primary topic of conversation.

Thomas Nurnberger of Northwestern Bell summed up his attitude to McGowan this way: "You bastards are not going to take away my business."

Charles Brown, the head of Illinois Bell, agreed that competition threatens "large amounts of revenues [that] we can preserve if we choke it off right now."

A year later, when Bill McGowan sued AT&T for violating the antitrust laws by trying to squash MCI, the notes from the Key Largo meeting found their way into court. MCI's antitrust lawsuit, which sought $900 million in damages, would take over a decade to litigate and appeal. In the meantime, McGowan began the next phase of his plan to trick the FCC into allowing competition into the communications market.

His opening came in 1973, courtesy of the new AT&T chief executive, John deButts. Angered by McGowan's temerity and success in manipulating the FCC, deButts publicly declared that the Bell systems would do what it took to restore its position as an absolute monopoly, even if that meant ignoring regulatory decisions it didn't agree with. Even coming from the phone company, the hubris shocked McGowan. It also offered an opportunity. Most of the staff members at the FCC were used to doing AT&T's bidding, but only while the phone company publicly maintained the fiction that the regulators were the ones in charge. The open intransigence of John deButts seemed to particularly offend Bernie Strassburg, a key FCC staffer watching the speech, McGowan noticed.

With AT&T was focused on MCI's antitrust lawsuit, McGowan approached Strassburg and pitched the idea of modifying the operating license the commission had granted to MCI. McGowan knew that if he simply asked the FCC to let MCI connect into the Bell system's local phone exchanges—exposing the market for long-distance phone service to its first-ever dose of competition—he would be voted down instantly. Instead, he hoped to win FCC authorization to sell long-distance telephone business without the FCC commissioners realizing what they had done. Two weeks after deButts's maddening speech, McGowan wrote a letter to Strassburg asking for "clarification and opinion regarding orders we have received from customers for specific types of interstate service."

"We would appreciate your confirming that we are entitled and obligated to supply these services. We have orders for these types of arrangements and are anxious to provide such services in accordance with our rights and obligations under the law."

Two days later, Strassburg took MCI's letter with him to a commission meeting and passed it off as just another of the countless mundane regulatory matters the commissioners were responsible for overseeing. For the rest of the day, more important matters than a minor rule tweak affecting a minor company occupied the commissioners' time. The following day, October 4, 1974, they approved it.

McGowan was careful to establish a regulatory record that would leave him in the strongest possible legal position when AT&T finally figured out what he had done. The wording of the letter approved by the commission had been intentionally vague, so McGowan wrote Strassburg a letter translating the authorization into technical language: "We understand that our letter, signed by the commission on October 4, authorizes us to sell FX lines," he wrote, referring to the abbreviation for long-distance telephone lines. Four days later, Strassburg sent a letter confirming that interpretation. A few weeks after that, before his colleagues realized what he had done, Strassburg retired.

This was the way to getting the FCC to support good public policy, Bill McGowan believed. Only a fool would walk into the regulatory agency, make a straightforward case for competition, and expect to win.

IT TOOK MCGOWAN'S ENEMIES at the phone company six months to realize that they had their first challenger in the long-distance business, and then no time at all to race to FCC headquarters to demand the competition be outlawed. Outraged AT&T representatives showed the regulators McGowan's trick; MCI customers used a touchtone phone to dial a local phone number that connected them to MCI's network, then typed in their twelve-digit customer ID, and finally entered the phone number they actually wanted to call in a different city. It took the FCC members several moments to grasp the implications of the demonstration. Then it dawned on them: Bill McGowan had broken AT&T's monopoly on long-distance phone service, kicking out a central pillar of federal communications policy—and he had done it without ever letting the FCC in on his plan.

All seven FCC commissioners and their staff members were livid. That MCI's tiny network had so far generated revenues of just $7 million (versus $27 billion at Bell system) did nothing to mollify them. Within weeks, Bill McGowan found himself holding a new decision from the FCC, ordering MCI to immediately stop selling long-distance phone calls.

As a legal justification for putting McGowan out of business, the FCC parroted the argument often made by AT&T. Its long-distance monopoly should not face competition, the Bell system argued, because the phone system needed the high profits from long distance to subsidize ordinary home phone lines. If the FCC didn't ban McGowan, AT&T chief John deButts warned, a few big businesses might save money on long distance, but regular citizens would see their home phone bills soar. McGowan, who knew how inefficiently the local phone networks were run, considered this a flat-out lie. A unanimous Federal Communications Commission accepted the argument unconditionally, so eager to run the scheming Bill McGowan out of business that it issued MCI's death sentence without even bothering to hold a hearing to consider the condemned company's side.

As McGowan paged through the FCC's hastily drafted order, the more reasons he found to smile. No doubt the lawyers at AT&T were smiling too, pleased with their apparent success at killing of another would-be competitor, unaware that they had blundered straight into Bill McGowan's trap.

IMMEDIATELY AFTER RECEIVING the FCC order running them out of the long-distance business, McGowan and his lawyers headed to federal court to sue the FCC. This time Federal Judge Skelly Wright took on the job of authoring a biting opinion explaining to the FCC's lawyers the merits of McGowan's legal argument. While the commission had the authority to grant AT&T a monopoly in long distance, the court ruled, McGowan was correct in pointing out that it had never formally done so. Before the FCC could banish MCI for breaking AT&T's legal monopoly in long distance, it would first need to go through a formal process to make the monopoly official. That prospect worried Bill McGowan not at all, since he could think up a dozen specious procedural objections his lawyers could use to gum up the FCC's bureaucracy indefinitely.

With McGowan racking up victories in court, AT&T executives unveiled a new plan to kill MCI. This time they deployed their ultimate weapon against new competitors: the United States Congress. If federal law prevented the FCC from banning competition, then AT&T would change the law, just as it had with the "The Satellite Communications Act of 1962." In 1976, the Bell system convinced a friendly congressman to draft a new law doing exactly that. Soon the "The Consumer Communications Reform Act" had over two hundred congressional cosponsors pushing for its passage.

Once again, Bill McGowan was one step ahead and responded with a well-planned counterattack. He mocked the sophistry of the law's name and succeeded in rebranding it in the mind of the media as "The Bell Bill." On Capitol Hill, MCI shamelessly adopted the argument traditionally used by the cartels. Congress shouldn't get involved in micromanaging the industry, the company told lawmakers, but instead rely on "the expert agencies which you and the president have created to deal with communications—and which have been dealing with or studying these issues for a long time."

Trust in the Federal Communications Commission to understand technology and do the right thing—of all the deceptive arguments Bill McGowan made during his epic fight with the Bell system, this was surely the most disingenuous of all.

Meanwhile, MCI began a grassroots lobbying effort based for the first time on the arguments McGowan actually believed in but had been too cagey to state openly. The logic for limiting competition in communications no longer applied, he pointed out, since digital technology made overlapping telephone networks affordable to build and easy to interconnect. McGowan's case for killing the bill—and doing away with the Bell cartel that supported it—won backing from across the political spectrum, from Ralph Nader to Milton Friedman.

The sort of honest arguments in favor of competition that invariably failed before the Federal Communications Commission could be highly successful if you could get the American people to pay attention, McGowan proved. Once incumbent congressmen started getting hit with attack ads painting them as the phone company's toadies, the number of cosponsors began to plummet. Representative Timothy Wirth, the chairman of the House Telecommunications Subcommittee, became the Bell Bill's fiercest opponent, declaring the stakes involved to be "nothing less than the control of information in a democratic society." Eventually, a humiliated AT&T had the bill pulled from consideration. The Bell Bill, McGowan would later gloat, "failed so miserably. All it did was attract the attention of a lot of forces on the Hill, who turned out to say, Now that we're studying this, why don't we do the opposite? 'Why don't we encourage competition and break up the Bell System?' "

AFTER ORDINARY AMERICANS won the right to choose between different long distance telephone companies, Bill McGowan found himself with one final role to play. As competition slashed the price of a long-distance phone call, new customers flooded his microwave network. By the early 1980s, many of McGowan's microwave towers had hit capacity, spurring him to consider a new technology that promised to offer nearly unlimited information-moving capacity.

In 1983, with the help of financier Michael Milken, McGowan sold $1.1 billion in convertible bonds, breaking the record for the largest single debt financing in history. With that money, MCI construction crews began digging trenches alongside Amtrak train tracks, dropping in cables containing slender glass fibers. McGowan called them fiber optic cables, not laser pipes, but the technology worked just as David Sarnoff had envisioned. MCI's new digital network, true to Sarnoff's vision, eliminated "any distinction among the various forms of communications [allowing all of them to] pass simultaneously through the same relays in the form of identical electronic pulses."

In a deal that would prove critical to making the rest of David Sarnoff's final prediction a reality, McGowan teamed up with the National Science Foundation to build a pioneering computer network, using his new laser pipes. To ship digital data around the country, the new network combined MCI's digital network and a new set of rules for efficiently routing the flow of information from one person to another. The name of that obscure standard–Internet protocol–would soon become known around the world.

The interest of the dealers in any particular branch of trade or manufactures, is always in some respects different from, and even opposite to, that of the public. To widen the market and to narrow the competition, is always the interest of the dealers.

To widen the market may frequently be agreeable enough to the interest of the public; but to narrow the competition must always be against it, and can serve only to enable the dealers, by raising their profits above what they naturally would be, to levy, for their own benefit, an absurd tax upon the rest of their fellow-citizens.

The proposal of any new law or regulation of commerce which comes from this order ought always to be listened to with great precaution, and ought never to be adopted till after having been long and carefully examined, not only with the most scrupulous, but with the most suspicious attention. It comes from an order of men whose interest is never exactly the same with that of the public, who have generally an interest to deceive and even to oppress the public, and who accordingly have, upon many occasions, both deceived and oppressed it.

—Adam Smith, The Wealth of Nations.

ACKNOWLEDGMENTS

The generosity and help of many people made this book possible.

Brian Bergstein, Jeff Garigliano, and Dan Roth offered early inspiration, criticism, and editing. Dennis Kneale, my longtime editor at Forbes and a man born to edit narrative nonfiction, put in long hours on early drafts of almost every chapter. I also benefited from being married to a brilliant writer who can mix unsparing criticism of flabby prose (sample margin note: "You are better than this") with unstinting support for a book project that lasted for a very long time. I love you, Bronwyn.

I owe special thanks to my agent Jay Mandel, in particular for his early willingness to believe that my tale of a secret conspiracy dedicated to controlling invisible waves in the air might be a book, not a delusion. Hilary Redmon and Emma Janaskie at Ecco displayed a similar open-mindedness, as well as providing valuable advice and invaluable patience as the narrative mutated from initial proposal to finished product.

Any talent I possess at pulling apart the economics of an industry I owe to my time studying under a number of brilliant microeconomists, in particular Jay Prag and Ross Eckert at Claremont McKenna and F. M. Scherer and Richard Zeckhauser at Harvard. I also owe a great debt to Watson Branch and Jack Pitney, two teachers who opened my eyes to the joys of attempting to write well.

Thanks to the unflinchingly honest Meredith Haberfeld for helping me turn a vague idea into an actual book proposal. Thanks to the brilliant and bold Carys Woolley, in particular for picking up were Meredith left off and encouraging me to actually finish. Thanks to the calm and cheerful Graham Woolley for reminding me that a book isn't the most important thing in my life. And a shout out to Buzz Woolley, who helped me learn a lesson about what it means to be a patient father.

I DEDICATE THIS BOOK to the memory of my wonderful mother, Ellen Omsted Woolley.

A BRIEF DISCUSSION OF SOURCES

The story told in *The Network* grew out of five years of original reporting, during which I collected thousands of primary sources to shape the main narrative. The most important of those sources include: the testimony of David Sarnoff, Edwin Armstrong, and others in Armstrong's 1948 lawsuit against RCA; the personal papers of Sarnoff, Armstrong, and Owen Young; the evidence collected during the House and Senate investigations into FM radio; company records of the Marconi Wireless Telegraph Company, RCA, AT&T, and NBC; and the records of the Federal Communications Commission.

A far more extensive list of primary sources can be found online at BattleForTheAirwaves.com, including direct links to hundreds of letters, diaries, transcripts, eulogies, photographs, and much more.

In three of the book's chapters, primary sources and original research ended up playing a supporting role. In those chapters, the story overlapped with a previous work of nonfiction that had been so comprehensively researched that adding new reporting became largely unnecessary and original research served mainly to verify specific facts and stitch pieces of those previous narratives into this book's larger story.

The chapter that focuses on Lyndon Johnson was built

around the work of three first-rate authors. Michael Kinsley's 1976 book, *Outer Space and Inner Sanctums,* offers a definitive account of AT&T's efforts to cripple the satellite industry in 1962. The sections on LBJ and Lady Bird's broadcast empire are based mainly on Robert Caro's *Means of Ascent* and Robert Dallek's *Lone Star Rising,* both excellent biographies. (The chapter's details on other congressmen's broadcasting investments comes mostly from FCC records.)

The epilogue's retelling of how Bill McGowan toppled AT&T traces a story told first (and in much greater detail) in Steve Coll's 1986 book, *The Deal of the Century: The Breakup of AT&T.* The epilogue also benefits from two sources that were lacking to Coll: the long and frank interview McGowan gave as part of an oral history project in 1988 and Bernard Strassburg's 1988 book, *A Slippery Slope: The Long Road to the Breakup of AT&T,* which offers an inside-the-FCC view of the MCI versus AT&T story.

The chapter focusing on Bud Rogers and the local broadcasters is based mainly on Rogers's own telling of the story in his candid autobiography, *History of Television, A Personal Reminiscence.* (Roger's account was also buttressed by and fact-checked against other accounts and sources.) That autobiography provides a tremendous public service, though perhaps not the one its author intended, by offering a firsthand look inside the local broadcasters' cartel.

The picture painted of Guglielmo Marconi—and the implication that he lied about hearing a transatlantic signal in 1901—is based on a combination of scientific and circumstantial evidence. While Marconi's claim to have heard a signal sent across the Atlantic was barely plausible to scientists in 1901, twenty-first-century science has only added more reasons to doubt his story. The one possible way that a signal could have crossed the Atlantic during the daytime, according to several modern radio propagation experts, was if (1) Marconi's transmitter in England, designed to broadcast waves 365 meters long, accidentally sent out waves just 40 meters long, and (2) His receivers, designed to catch the much longer waves, also happened to mistakenly tune into the much shorter waves. While Marconi made it impossible for his doubters to disprove what he did or did not hear in his headphones, his motive to lie in order to keep his company solvent combined with the scientific implausibility of his claim is enough to convict him of scientific fraud.

The book's portrait of the FCC as a timid and feckless regulator builds on the work of Ronald Coase, an economist who would go on to become a Nobel laureate, and who described the agency in 1966 as being so hopelessly in the thrall of the companies it was supposed to regulate that it "cannot conceive of a future which is not essentially a repetition of the past." Vincent Mosco's 1975 book, *Broadcasting in the United States: Innovative Challenge and Organizational Control,* further explored the FCC's many failings. Thomas Hazlett, who served as chief economist for the FCC, has written several critiques of his former employer.

Owen Young's extensive papers, available at the library of St. Lawrence University, provided invaluable records from the early days of RCA. The Herbert Hoover Library in West Branch, Iowa, has the best records of Commerce Secretary Hoover's series of radio conferences, including the 1924 conference that is the focus of Chapter Seven. Sarnoff's speeches are available from several sources. The David Sarnoff Library, once located in Princeton, has closed and transferred it materials to the Hagley Library in Delaware, which is working to digitize much of the collection. (Alexander Magoun, former curater of the Sarnoff library, also wrote *Television: The Life Story of a Technology*, which offers a comprehensive history of TV's birth.) Edwin Armstrong's collection of personal papers, available at the Columbia University Rare Book & Manuscript Library, also offered a trove of valuable information.

David Sarnoff's remarkable speech predicting the rise of fiber optics and the Internet was made in 1965, but has been ignored until now.

AN EXTENDED SELECTION OF IMPORTANT SOURCES BY TOPIC

GENERAL ARCHIVES

Edwin H. Armstrong's papers, Columbia University Rare Book & Manuscript Library.

David Sarnoff Archives, Hagley Museum & Library.

Owen D. Young Papers, St. Lawrence University.

The Marconi Company Archives, Bodleian Library, University of Oxford.

Herbert Hoover's papers, Hoover Library, West Branch, Iowa.

The Papers of Clay T. Whitehead, (1938–2008). Available in the Library of Congress.

ECONOMICS

The Federal Communications Commission, by Ronald Coase. *Journal of Law and Economics* 2 (October 1959).

"A Property System for Market Allocation of Electromagnetic Spectrum: A Legal-Economic-Engineering Study," by Arthur S. de Vany, Ross D. Eckert, et al. *Stanford Law Review* 21:6 (June 1969).

Broadcasting in the United States, by Vincent Mosco. Ablex Publishing Corp., 1979.

WIRELESS TECHNOLOGY & COMMUNICATIONS THEORY

Early FM Radio: Incremental Technology in Twentieth-Century America, by Gary L. Frost. The Johns Hopkins University Press, 2010.

The Continuous Wave: Technology and American Radio 1900–1932, by Hugh G. J. Aitken. Princeton University Press, 1985.

Wireless: From Marconi's Black-Box to the Audion, by Sungook Hong. The MIT Press, 2010.

The Transistor, A Semi-Conductor Triode, by J. Bardeen and W. H. Brattain. *Phys. Rev.* 74:230 (July 15, 1948).

Petition of Radio Corporation of America and National Broadcasting Company, Inc., for approval of color standards for the RCA color television system. Before the Federal Communications Commission, 1953.

A Mathematical Theory of Communication, by Claude Shannon. *Bell System Technical Journal* 27 (July, October 1948), pp. 379–423, 623–656.

REGULATORY HISTORY & THEORY

Recommendations for Regulation of Radio Adopted by the Third National Radio Conference, October 6–10, 1924. Issued by the Government Printing Office.

To Regulate Radio Communications. Hearings before the U.S. House of Representatives, Committee on Merchant Marine and Fisheries, March 1924.

Records of the Select Committee to Investigate the Federal

Communications Commission. U.S. House of Representatives, 1944.

Small business opportunities in FM broadcasting: report of the Special Committee to Study Problems of American Small Business, U.S. Senate, 1946.

Radio and Television Regulation: Broadcast Technology in the United States, 1920–1960, by Hugh R. Slotten. The Johns Hopkins University Press, 2000.

The Red Scare, Politics, and the Federal Communications Commission, 1941–1960, by Susan Brinson. Praeger Publishers, 2004.

Progress of FM Radio. Hearings of the U.S. Senate Committee on Interstate and Foreign Commerce, March 1948.

Investigation of Regulatory Commissions and Agencies. U.S. House of Representative, 85th Congress. Hearings before a Subcommittee of the Committee on Interstate and Foreign Commerce, 1958.

The Crisis of the Regulatory Commissions, Paul MacAvoy (editor). W. W. Norton, 1970.

Cable Television and the FCC: A Crisis in Media Control, by Don R. LeDuc. Temple University Press, 1973.

Federal Regulation and Regulatory Reform. Report by the Subcommittee on Oversight and Investigations. House of Representatives, 94th Congress, October 1976.

CONGRESSIONAL CORRUPTION

The Professor & the Commissions, by Bernard Schwartz. Alfred A. Knopf, 1959.

Outer Space and Inner Sanctums, by Michael Kinsley. John Wiley & Sons, 1976.

Means of Ascent: The Years of Lyndon Johnson, by Robert Caro. Random House, 1990.

Lone Star Rising: Vol. 1: *Lyndon Johnson and His Times, 1908–1960,* by Robert Dallek. Oxford University Press, 1991.

COMPANY & INDUSTRY INFORMATION

A History of the Marconi Company, by W. J. Baker. Methuen, 1970.

John Fletcher Moulton and Guglielmo Marconi: Bridging Science, Law and Industry, by Anna Guagnini. Notes & Records of the Royal Society, 2009.

Directors' Report of the Marconi Wireless Telegraph Company, various 1900–1915.

Report of the Federal Trade Commission on the Radio Industry, 1923.

RCA Annual Reports, 1920–1965.

AT&T Annual Reports, 1912–1975.

FCC Annual Reports, 1934–1971

Interviews with Bill McGowan, The MCI History Project, 1988.

The History of MCI, 1968–1988, by Philip L. Cantelon. MCI Communications Corp., 1993.

BIOGRAPHIES

The General, by Kenneth Bilby. Harper & Row, 1986.

Owen D. Young and American Enterprise, by Josephine Young Case and Everett Needham Case. David R. Godine, 1982.

Beating the Odds: The Untold Story behind the Rise of ABC, by Leonard Goldenson and Marvin J. Wolf. Scribner, 1991.

Cover Story on David Sarnoff, *Time* magazine, July 23, 1951.

LAWS, LEGAL CASES, & LAW REVIEW ARTICLES

Edwin H. Armstrong v. Radio Corporation of America and National Broadcasting Company, civil action no. 1139. United States District Court for the District of Delaware, 1948.

National Broadcasting Company, Inc., et al. v. United States, 319 US 190. Decided by Supreme Court on May 10, 1943.

MCI Telecommunications Corporation v. Federal Communications Commission, 561 F. 2d 365 (D.C. Cir.), 1977.

Hush-A-Phone v. United States, 238 F. 2d 266 (D.C. Cir.), 1956.

WHDH: The FCC and Broadcasting License Renewals, by Louis L. Jaffe. 82 *Harvard Law Review* 1693 (1968–1969)

"Ownership of Broadcasting Frequencies: A Review," by Paul M. Segal and Harry P. Warner.

Fortnightly Corp. v. United Artists Television, Inc., 392 US 390, 1968.

MISCELLANEOUS

The Eulogy of Edwin H. Armstrong, by Thornton Penfield Jr. (Text of original sermon courtesy of Reverend Penfield's family.)

Architectural Diagrams of River House and Edwin Armstrong's Apartment. (Courtesy of Columbia University School of Architecture.)

Titanic Disaster Hearings: the Official Transcripts of the 1912 Senate Investigation.

Investigation of the British Wreck Commissioner, 1912. Right Honourable Lord Mersey, Wreck Commissioner.

Video of the 1921 Dempsey-Carpentier title fight (widely available online).

Reports from the Select Committee on Marconi's Wireless Telegraph Company, Limited, Agreement (House of Commons, 1913).

Most Secret War, by R. V. Jones. Hamish Hamilton, 1978.

National Centers for Environmental Information, Center for Weather and Climate.

Broadcasting Magazine, various 1931–1972.

R.C.A. Review, various, 1936–1962.